COMETS: VAGABONDS OF SPACE

Comet Tago-Sato-Kosaka (1969 IX), Jan. 1, 1970, 0915 hrs. U.T. (Universal Time), as photographed by Mr. Barrie Ward of Tirau, New Zealand. 300-mm telephoto lens on 35-mm camera. Exposure time 3 minutes on Kodak Tri-X film.

COMETS:
Vagabonds of Space

DAVID A. SEARGENT

1982
Doubleday & Company, Inc., Garden City, New York

Library of Congress Cataloging in Publication Data

Seargent, David A.
Comets: vagabonds of space.

Includes index.
1. Comets—Popular works. I. Title.
QB721.4.S4 523.6
AACR2

ISBN 0-385-17869-7
Library of Congress Catalog Card Number 81–43636

Printed in the United States of America
First Edition

TO MY PARENTS, GRANDFATHER,
AND TO THE MEMORY OF MY GRANDMOTHER.

ACKNOWLEDGMENTS

In the writing of a book such as this, assistance inevitably comes from many different people in many different ways. A word of encouragement or criticism here, a correction there, a suggestion or a word of advice.

Naturally, it is practically impossible to recall everyone who has ever (directly or indirectly) exercised influence upon the writing of this book—their number is surely legion.

Nevertheless, to those whose services were especially significant, let me extend my deepest thanks. To Reinder Bouma and Peter Bus of the Dutch Comet Section, Drs. David Meisel, F. D. Miller, H. L. Giclas, M. J. Bester, and Mr. B. Ward for their permission to use the photographs included in this volume; to Mr. Spencer Vale and Mrs. Glenda Griesberg for their help in the preparation of some of the photographs and sections of the manuscript, to the California Institute of Technology for supplying several fine photographs, and to the Tokyo Observatory for their photographs of Comet Ikeya-Seki (1965 VIII) under most unusual conditions.

D.S.
The Entrance
New South Wales, Australia
1981

CONTENTS

PREFACE

A fully developed comet with a strong tail of considerable length is one of nature's finest sights, one which the average person has a greater chance of seeing than those other spectacles of equivalent magnificence, such as great meteor showers, total solar eclipses, or volcanic eruptions. An owner of a wide-field telescope or a large pair of binoculars may, indeed, hope to see such a sight every few years, but for the majority of citizens, the chance comes much more rarely. Naked-eye comets are frequent enough, but few are seen in their full splendor without optical aid. Only once every decade or so does one become sufficiently bright to be conspicuous. Nevertheless, several "great comets" may be expected to appear during the lifetime of the average person—and they are certainly well worth the wait!

Since 1960 there have been three comets which could be termed "great" (in 1965, 1970, and 1976), plus a fourth which looked promising for a time and attracted wide attention, despite a rather disappointing final performance—Kohoutek (1973 XII). In addition, the same period saw a number of lesser—though still conspicuous—naked-eye objects, including the "fairly great" Seki-Lines (1962 III), Wilson-Hubbard (1961 V), Tago-Sato-Kosaka

(1969 IX), and White-Ortiz-Bolelli (1970 VI), as well as others of lesser prominence, like Ikeya's 1963 I.*

These recent events—plus the rising interest in comets among professional astronomers as new and powerful methods of observation are employed in their study—have led to increasing interest in these strange objects among the general public. This will probably increase rather than diminish in coming years as the 1986 apparition of Halley's Comet draws closer.

One measure of public interest is the number of books now being devoted to the subject. When I first became seriously interested in the topic (in the early 1960s), books devoted solely to the study of comets were rare and almost unprocurable, even at libraries and specialist book shops. Ten years later, I was encountering elementary comet books on the newsstands!

At the other end of the spectrum, more space in scientific publications is being devoted to the technical discussions of professional astrophysicists, centering upon various aspects of cometary physics, and an observatory constructed especially for cometary research is now operating at South Baldy in New Mexico.

This book is intended to lie somewhere in the middle of the spectrum. It is the result both of a long interest in comets and several years of research into what I hoped would eventuate as a comet book of interest to amateur astronomers who, having received a smattering of knowledge on the subject, had their appetites whetted for more information, not of the technical, "head-hurting" variety but of the type likely to be of interest to people rather like myself. Consequently, this work is "subjective" to a certain degree in that it reflects my own interests in the field and represents an attempt to present amateurs and other interested persons with the type of book for which I longed in my early days. Trusting my own psychology not to be unique among amateur astronomers, I further trust that it will (however imperfectly) fulfill just such a need.

D.S.
1981

* The naming of comets, a complicated matter, is explained in detail in Chapter 2. Essentially, they are denoted by the name (or names) of the discoverers (up to three), the year of perihelion passage, followed by the order (roman numeral) in which that comet passed perihelion that year. New comets are initially denoted by the year of discovery, followed by a, b (first, second), etc.

INTRODUCTION

Perhaps no other phenomenon of nature—certainly no other astronomical display—has caused more panic and awe than the phenomenon of the comet. Volcanos, earthquakes, tornados, and the like—even eclipses and meteor showers—may appear more terrifying, but they are both localized and brief. Comets, on the other hand, become visible over much of the Earth, hanging in the sky for several nights—maybe even for weeks or (on rare occasions) months—suspended like the Sword of Doom itself.

Differing both in appearance and in their manner of motion from anything else in the heavens, they appear suddenly, move mysteriously (or so it seemed to the ancients), and then fade into oblivion as if coming to arouse some dreadful calamity and then, their mission completed, to retire once more into the darkness from whence they came.

From the very dawn of human intellect, men must have speculated on the nature of these mysterious wanderers of the night. At first, undoubtedly, this speculation was wholly inspired by fear, but gradually a more objective and scientific approach began to emerge.

The Babylonians, for instance, subscribed to the surprisingly modern belief that comets are strange types of planets. This view, however, was not widely held among ancient scholars and was not generally believed until quite modern times.

On the other hand, the Greek historian Ephorus was of the opinion that comets resulted from the combination of two or more stars, and he cited as evidence the division of the great comet of 371 B.C. Other Greeks shared this (or similar) opinions, although this general line of thinking was attacked by Aristotle, who put forward the counter idea that comets are "exhalations" thrown up from the Earth and made luminous on reaching the sphere of fire which was supposed to encircle the world.

It is interesting to note that while Aristotle has preserved for us a reasonably detailed account of the great comet of 371 B.C., he makes no allusion to its partition, which raises the suspicion as to whether Ephorus may have been guilty of a little embroidery in support of his theory, or alternatively whether Aristotle may have "overlooked" an observation difficult to explain by his thesis!

Incidentally, many people now believe that the comet of 371 B.C. may have passed very close to the Sun and, consequently, some disruption may have occurred. The account of Ephorus may be correct after all, although it remains true that the disruption as described by him would have involved a much more violent division than what has been noted for recent "Sun-grazing" comets.

Be this as it may, "tectonic" (geophysical) and "meteorological" explanations, such as that of Aristotle and those which relegated comets to a position similar to rainbows, held precedence over astronomical ones in early times. This is not really surprising upon reflection. Bright comets with long tails would have been, of course, the ones mostly observed by early astronomers, and these comets (for reasons which will become apparent in Chapter 1) tend to be located near the Sun—they are observed either in the west after sunset or in the east before dawn, with their tails directed "up" into the sky. No great imagination is needed to interpret such a phenomenon as a reflection of the Sun or a rush of gas from a volcanolike disturbance somewhere on Earth.

Astronomers, then, largely disowned comets. Pliny the Elder believed them to be meteorological phenomena—as apparently did Ptolemy, if their omission from the *Almagest* is any indication of what he thought of them. Even in medieval Europe, when comets were not actually being looked upon as portents of coming disaster (the usual interpretation at the time), the meteorological hypothesis held sway. Even Galileo believed them to be mere creations of sunlight, rather like rainbows and equally as insubstantial, and his theories regarding this matter were quite well received at the time.

Like the Occidentals, the people of the East (especially the Japanese) regarded comets with superstitious terror in ancient and medieval times. It is difficult to determine exactly what the Orientals believed comets to be, although at least one kind seems to have been viewed as a celestial broom with which the gods swept Heaven free of evil—casting it (naturally) down upon poor old Earth! Even today this superstition seems to persist in certain parts of Asia, as is evidenced by the fear aroused among Vietnamese peasants in 1970 at the appearance of a "Sky Broom"—presumably the great comet of that year, known more prosaically in the West as Bennett's Comet (1970 II). The opinion was quite widely held that war in their land would increase and that generally bad times lay ahead. (Incidentally, this same comet was actually shot at by troops in the Middle East—they thought it was a secret weapon!)

The modern conception of comets did not really begin to emerge until the late sixteenth and seventeenth centuries. Tycho, in 1577, demonstrated that the great daylight comet of that year was more remote than the Moon and in consequence a genuinely astronomical object. Accurate measurements of its apparent celestial position revealed a negligible parallax when taken from Hven (in Sweden) and Prague, thus furnishing the first proof that comets are truly celestial. Furthermore, Kepler repeated Tycho's experiment with the comets of 1607 and 1618, obtaining the same result.

Meteorological and other "terrestrial" theories were now gone forever, but problems certainly had *not* gone forever. If comets are astronomical objects, why are they so different from the rest? They seemed to be made of a different stuff than the stars and planets are, and even their motion across our sky seemed to run counter to the relatively orderly motion of the planets.

Tycho thought comets moved in circles; Kepler, in straight lines. It remained for Edmond (not Edmund) Halley (pronounced, according to Ronan, "Hawley"; "Hall" as in "town hall", "ey" not to rhyme with "valley" or "Bailey") to discover the true nature of cometary orbits.

In the year 1680 a great comet passed very close to the Sun, and two years later another swept into view. Halley suspected a connection between both these objects and others observed in the past. Could it be possible, he came to ask, that there is such a thing as a *returning* or *periodic* comet? Truly such questioning was very revolutionary for Halley's day.

The first comet, Halley mused, might be a return of the great daylight comets of 1106 A.D. and 44 B.C. (the comet which appeared at the time of the death of Julius Caesar and which is immortalized by Shakespeare in the lines: "When beggars die there are no comets seen. The heavens themselves blaze forth the death of princes."). The year 531 A.D. also saw the appearance of a substantial comet, and Halley reasoned that a single object having a period of about 575 years may have accounted for all three apparitions, in addition to the great comet seen by himself in 1680.

Similarly, could the great comet of 1682 be a reappearance of those seen in 1607 and 1531? If this comet was indeed periodic, it should be moving in an ellipse with a period of only 75–76 years and the suggestion should be open to verification—a great comet should appear in 1758, or thereabout.

The rest of the story is well-known history. The comet was recovered (resighted) on Christmas night 1758 and sailed into full view early the following year.

Halley's other comet has not, incidentally, been so fortunate. It has since been proved that the comets of 44 B.C., 531 A.D., 1106, and 1680 were separate objects and not the returns of a single periodic object.*

Other periodic comets have been discovered since Halley's day, however, and in recent years new ones are being added to the list almost annually, although most of these are very faint.

A notable early example is Encke's Comet, with a period of only three years and four months. This is the first comet of the

* Ironically, the comet of 531 A.D. may actually have been Halley's Comet itself. Halley's Comet returned in 530, and the rather vague records of a comet the following year may refer to the extended visibility of this object.

central short-period group to have been discovered and even to this day remains the one with the shortest well-established period.

As we move into the twentieth century, the status of comets undergoes a revolutionary change. No longer are they mysterious portents in the sky—they obey the same natural laws as all other material bodies moving freely in space. No longer are they seen as rare and spectacular objects—in fact new telescopic comets are now discovered at the rate of at least one every couple of months, and it is seldom that at least one comet is not within the reach of some telescope somewhere. Mostly these telescopic comets are just little blobs of diffuse light drifting very slowly from one night to the next across the starry field of wide-angle telescopes.

The suppositions of Kepler (who stated that there were more comets in the sky than fish in the ocean) and Halley (who believed that a small and otherwise unobserved and unknown comet found by himself in 1717 was representative of a multitude of similar inconspicuous objects) seemed verified before the close of the nineteenth century.

Only the idea that comets are fiery and responsible for hot weather remained fairly popular. Even a well-respected scientific magazine published in 1880 carried a report from Australia concerning the great comet of that year in which a note of the unusual heat during the time of observation was included. As this report was made during the height of an Australian summer, I don't really think the observers needed to go as far as the comet to discover the source of their discomfort!

The constant stream of faint comets, in cold weather as well as hot, finally put an end to even this bit of romantic comet lore (so thoroughly, it would seem, that a few people laid the blame for an unusually *cold* and wet summer of 1973–74 at the feet of Kohoutek's Comet!).

Recent years have witnessed an explosion of research and discovery in all branches of astronomy, and comet study has not been excepted. Artificial satellites and space probes have uncovered hitherto unknown constituents among the cometary gases, new and more refined Earth-based observing techniques have shed additional light upon the sizes of solid particles emitted by these objects, and fast electronic computers have enabled orbital calculations to reach a high standard of precision. Furthermore, astronomers have come to realize that in comets we may be seeing actual portions of the original solar nebula, existing in a relatively unaltered state, and that the study of these objects might answer fundamental questions about the origin and evolution of the Sun and planets. This prospect alone has made the study of comets appear worthwhile.

Thus with years of superstition, pseudoscience, and neglect behind it, cometary research has at last ceased to be a stagnant backwater of astronomical study and has taken its rightful place in the march against ignorance. Let us now look more closely at the strange and fascinating objects this research is revealing for us.

REFERENCE

1. Colin A. Ronan. *Edmond Halley: Genius in Eclipse.* London: Macdonald, 1970, pp. 212–13.

COMETS: VAGABONDS OF SPACE

1

Anatomy of Comets

It is early morning. The sky is like black diamond-studded velvet, stained only by the ghostly river of the Milky Way and the wraithlike cone of the zodiacal light.

Suddenly you notice something which you have not seen before: in the east, near the zodiacal light, a faint shaft of luminosity like a flashlight beam in a smoke-filled room. Your attention is drawn to this mysterious apparition. It is not moving in front of the stars as would a distant searchlight beam, but as you watch, you detect the diurnal motion it shares with the stars.

More and still more of this faint ribbon of light appears from beyond the horizon, and you begin to notice something else about it. It is becoming narrower toward the horizon; also, it is not perfectly straight as you initially thought, but very slightly curved.

Then, as the first rays of dawn light the eastern horizon, you see the termination of this strange curving light shaft. It is like a star—not a particularly bright star, but one which looks as though it is shining through thin clouds.

As you watch it rise higher into the heavens, the grand length of the object becomes fully displayed, before fading into the brilliant colors of the dawn.

Next morning you rise early and again seek this cosmic wraith.

There it is again! But could it be a little fainter this morning?

One thing certainly *has* changed. Yesterday it was below Betelgeuse; today it is noticeably above it and to the right.

Morning after morning you watch this ghost of space as it moves ever higher into the sky, fading all the while.

After about a week it has grown very dim. It would barely be visible now, but you know exactly where to look. Not only has it become dimmer, but the shaft of light is not so long now, not as curved and even more ghostlike, terminating now not in a starlike object but in a nebulous spot of definite, though small, size.

Finally, one morning you rise and seek your object, but find only the myriad stars of the predawn sky.

The above is not an account of any particular comet, though our hypothetical one is composed of features more or less common to the majority of these strange objects, though (let it be noted and always remembered) in conspicuously different degrees.

Neither, of course, should this hypothetical comet be looked upon as the average naked-eye comet. Average comets, like average anything-elses, exist (if at all!) only on paper. Indeed, it is a well-known fact that comets are more resistant than most to the pressures of being "average"—it is said that at a recent symposium, someone asked, "What is a normal comet?" Predictably, no answer was recorded.

Seldom are two comets closely alike. In fact, the range of magnitudes, appearances, and behavior is such that it is often difficult to believe that two comets are even members of the same *class* of objects. Who would guess, for example, that the tiny, faint, starlike point of light that was Comet Arend-Rigaux of 1957 could be the same type of object as the awe-inspiring Ikeya-Seki Comet of 1965, visible in daylight on the limb of the Sun and later spanning a quarter of the dome of the heavens?

How, too, could one believe that the little misty "star," weakly recorded on photographs in August and September 1909, would evolve into the great spectacle of May and June 1910?

Nevertheless, there are certain features which are, more or less, common to the great range of comets, although not necessarily all are present at any one time or in any particular object.

Figure 1 represents these various features. Figure 1a includes the more or less usual features, Figure 1b those which are more likely to be seen in large comets or otherwise active objects.

The most common features are the coma* and central condensation—the latter presumably masking the nucleus, which is invisible in many comets and is often difficult to properly distinguish from the central condensation in many others. Frequently—but certainly not inevitably—there is some trace at least of that feature for which comets are mostly noted: the tail.

We speak of a comet as being complete when it displays a nucleus, central condensation, coma, and tail. When there is a tail, the coma, central condensation, and nucleus together are termed the "head" of the comet, although for most purposes the expressions "coma" and "head" are taken as virtual synonyms.

We shall now look more closely at these major parts of comets.

THE NUCLEUS

When a bright comet is examined under high magnification and with the aid of a large telescope, its appearance may best be described as that of a star embedded in luminous mist. It is with this "star" that we are now concerned.

Generally, no amount of magnification will show it as anything more than a misty point of light or, at very best, a tiny planetlike disk. However, information as to its nature (like information regarding the nature of stars) can be gleaned by means of the spectroscope and also by careful monitoring of its brightness as the comet moves toward and recedes from the Sun.

The first important discovery revealed by spectroscopy is that the light by which we see a comet's nucleus is reflected sunlight.

* From a Greek word meaning "hair."

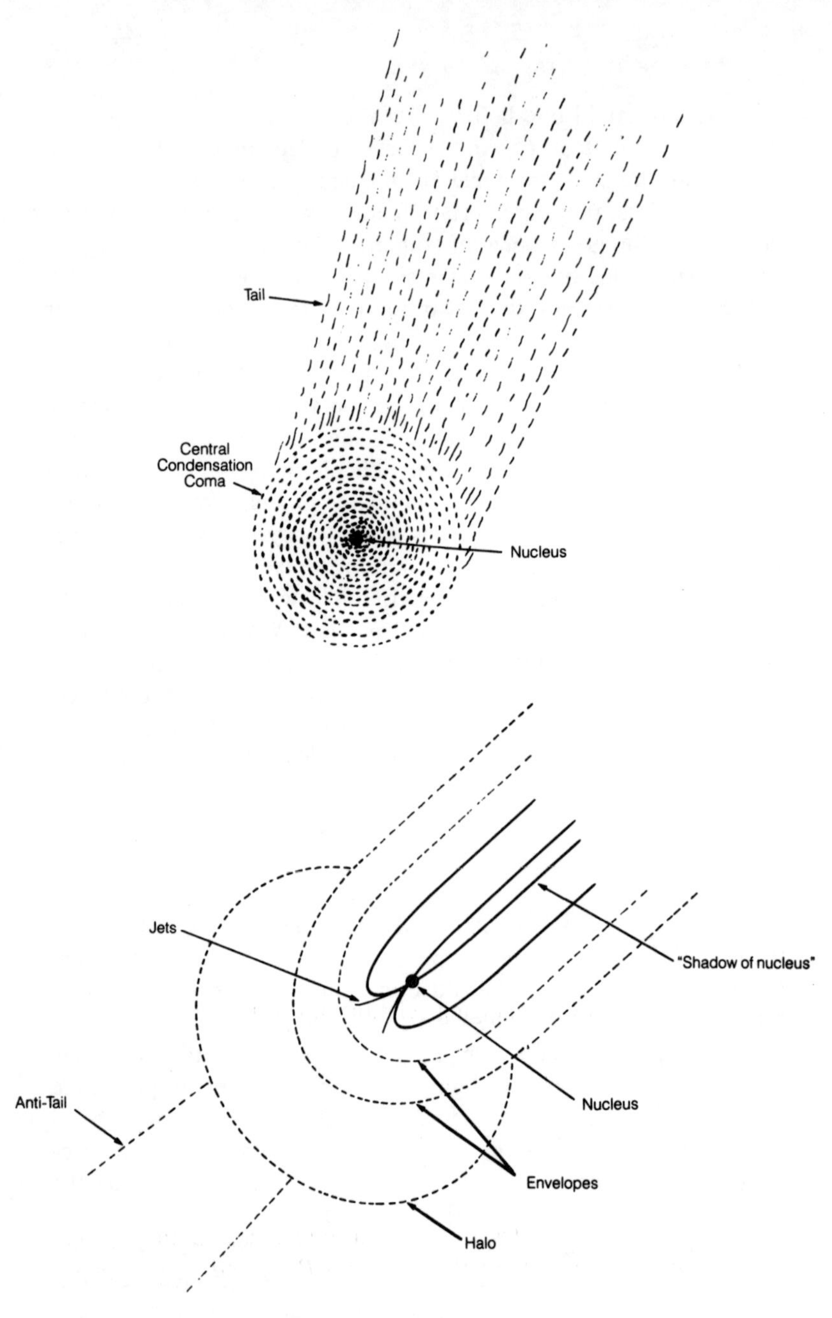

Fig. 1a. and 1b. *Diagrammatic representation of the parts of comets.*

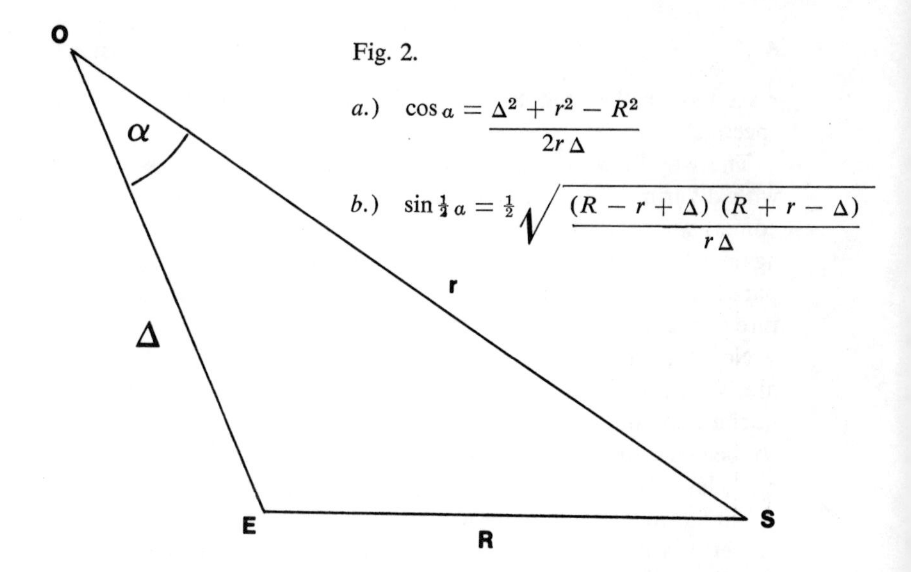

Thus, although it may have a superficially starlike appearance, it is certainly not starlike in nature but shines (with everything else in the solar system) only by the reflected glory of the Sun.

Now, knowing that an object shines by reflected sunlight enables other important information to be gleaned about its nature. Obviously, if an object reflects sunlight it must be a reflector; this much stands to reason! But there are reflectors and reflectors, and if we know the *kind* of reflector with which we are dealing, we shall be able to infer much about its nature as a physical object. How better to learn the nature of a reflector than by studying its reflected light?

To this end, it is important to note the way in which the magnitude changes with the changing relationship of the Sun and Earth —i.e., with changing *phase angle*.

This term will become clearer if we take a look at Figure 2.

If *E, S,* and *O* denote the Earth, Sun, and the astronomical object reflecting the Sun's light, α is the phase angle and may be determined by employing either of the formulae shown in Figure 2, the first of which will be recognized by all who have elementary knowledge of trigonometry as being an application of the "cosine rule" formula. The units of the distances *R, r,* and Δ are always given in astronomical units (A.U.), the main unit of distance in solar system studies. One A.U. is the mean distance of the Earth from the Sun and is equivalent to 149,600,000 kilometers. *R* and *r* are termed the "radius vectors" of the Earth and the comet, respectively.

The way in which the brightness of a comet's nucleus behaves with differing phase angles can be determined for any particular comet, and the corresponding phase curve (brightness graphed against the phase angle) can be constructed. If this is then compared with other phase curves of various types of objects, the nature of the nucleus may be found.

Now, the interesting results to emerge from such an exercise are, first, the obvious dissimilarity between phase curves of comet nuclei and solid bodies and, secondly, the rather striking *similarity* between these phase curves and those of clouds of suspended particles and bodies of dust such as the rings of Saturn.

Such evidence led many to the conclusion that the nucleus of a comet was *nothing but* a loose cloud of particles (meteoroids)—the diameter of the cloud being less than 50 kilometers in the vast majority of cases. Some authorities (most notably Dr. R. A. Lyttleton, whom we shall meet in Chapter 2) have even argued against the very notion of a nucleus, maintaining instead that the entire head of a comet is simply an attenuated cloud of very small dust particles and the starlike point at the middle is merely an optical effect due to the fact that the maximum depth of sight is through that region. While some line-of-sight effect must certainly occur, it is difficult to apply this thesis to those comets which have bright nuclei but only very faint comae or to the recent observations of Encke's Comet at the part of its orbit farthest from the sun. *Only* the nucleus of this comet is visible at these times and it is completely asteroidal in appearance—certainly not a line-of-sight effect.

Nevertheless, phase curve evidence does not prove that the nucleus is nothing but a cloud of dust or meteoroids. It only proves that dust and meteoric particles are present in rather vast numbers at the nucleus of a comet—but this is certainly not to say that there cannot be anything else there as well. A small fraction of the reflected sunlight could come from a solid body or from electron clouds (say), but wherever the most light was contributed by the particle cloud, the contribution from these other sources of reflection (if they exist) would be drowned out and fail to affect substantially the derived phase curve.

Indeed, various observations and theoretical considerations have increasingly led some astronomers to believe that a sizable solid body (or maybe, in some comets, several such bodies) exists within a comet's nucleus.

Dr. Fred L. Whipple is one such astronomer. In 1950, he proposed a model which has accounted so well for most cometary phenomena that it is now fair to term it the "accepted" or "orthodox" view.

According to Whipple, a comet's nucleus consists of an icy conglomerate: a mixture of water-ice, frozen gases, and large numbers of small meteoric "stones" and dust. As this mass (which we may suppose to average about 4 kilometers in diameter, although some are much larger and not a few considerably smaller) approaches the Sun, increased solar heating causes evaporation of the ices (in the vacuum of space, even water-ice evaporates rather than melts as the melting point of ice and boiling point of water are the same in a vacuum) with consequent freeing of meteoric stones and dust which were formerly embedded in the matrix.

Evaporating ices give these particles their required "lift," but many undoubtedly leave at such low velocities that they tend to travel along with the nucleus and, maybe, even orbit around it. In this way we can form the picture of a shrinking conglomerate slowly evolving into a more or less temporary cloud of meteoric particles—in good agreement with phase curve analysis.

Whipple assumes the ice mixture to contain the molecules CH_4, CO_2, NH_3, C_2N_2, and H_2O (methane, carbon dioxide, ammonia, cyanogen, and water) in addition to other highly complex molecules. From these "parent molecules" the observed "daughter molecules" are derived by the process of photodissociation. As we shall see in our subsequent discussion of the cometary coma, such a scenario is quite consistent with the changes actually observed to take place.

Moreover, many features of cometary behavior, such as jets and envelopes within the coma, are readily explained on this model. Thus as the cometary nucleus heats up on approaching the Sun, it is reasonable to expect certain "hot spots" to develop on the surface of the matrix—for example, in the vicinity of large dark boulders which absorb more sunlight than the surrounding area. Such hot spots will be expected to result in pockets of gas which (if sufficient pressure is built up) may burst forth in fountainlike eruptions, sending thin streamers of gas and dust far above the nucleus and curving back into the tail. The same effect would also result from the rapid evaporation of small pockets of volatiles in areas where any insulating surface crust which may exist is thinner or for some other reason less effective than normal. Such a crust may well be expected to form during the comet's journey through the dusty lanes of the solar system, and pieces of it may well break away, enabling intense irradiation of a single spot on the comet's surface to take place. It is reasonable to expect that some of the brightness flares associated with comets may also be explained in this general manner.

An interesting case which, in my opinion at least, lends strong support to the Whipple theory is provided by the behavior of Comet Tago-Sato-Kosaka (1969 IX) early in February of 1970. On the seventh of that month, a strong jet was observed both visually and photographically within the inner regions of the coma and quite clearly issuing from the nucleus.[1] The brightness of the comet started to increase anomalously, and a sudden increase in the thermal flux (electromagnetic variation) was also noted. For the next few days, the comet was judged to be about 1.5 magnitudes brighter than had been predicted, and the central condensation appeared abnormally sharp. The "flare" gradually diminished and the comet's brightness returned to normal, but some time later observers found the nucleus to be distinctly double, with the two components separating in such a manner that the actual moment of schism was calculated to have taken place on about the ninth of February—in other words, during the period of intense activity within the central condensation.[2] Clearly, the jet, the outburst, and the schism of the nucleus were all very closely related and point strongly to the presence of a solid icy object at the nucleus of this comet. In fact, such behavior as this is very difficult (if not outright impossible) to explain by a comet model consisting only of a swarm of dust or meteoroids.

In addition to the above considerations, some other observations should be noted which seem to suggest that there is something about a comet which is more solid than a swarm of meteoroids.

First, there are the very rare instances of the complete occultation of stars. Traditionally, when a star is occulted by a comet's head, there is little diminution in the star's light, although a certain amount of flickering and reddening has been occasionally reported. However, Chambers reported that in 1890 a star was completely obliterated by the nucleus of a comet. If we are to accept this observation at face value, it at least suggests either a solid body comparable in size to a small minor planet or a very dense cloud of meteoroids. Such a cloud would, however, be unlikely to remain stable for very long and would either collapse (in which case we arrive back at a solid object) or else disperse to such an extent that it would become transparent.

Of course, we are tempted to place an observation recorded as long ago as 1890 in the category of those nineteenth-century "observations" of vast atmospheres on Ceres and other asteroids, or volcanos on the Moon. Nevertheless, we should also remember that Bessel's observations of "wagging" in the tail of Halley's Comet in 1835 were rejected for well over a century—until exactly the same phenomenon was *photographed* in 1960 as taking place in the tail of Burnham's Comet (1960 II). Perhaps some modern astronomer will observe a complete occultation by a comet's nucleus, and although such events must be very rare, one should always keep watch for the possibility—paying especial attention to comets which cross rich star fields. A carefully calculated orbit could give advance notice of occultations or of appulses (i.e., close approaches to stars) which are likely to cause occultations, and the duration of such an event would certainly supply valuable data concerning the size of the nucleus.

Accurate measurements of stars occulted *near* the region of a cometary nucleus have already been carried out by Dossin for Burnham's Comet (1960 II) and similar observations formed part of the program for Kohoutek's Comet (1973 XII). Perhaps we will not have to wait long for an observation of a total occultation.

Closely related to this phenomenon are the transits of comets across the face of the Sun. An unconfirmed report of such an event in 1819 (Tralles' Comet) is not generally accepted, and nothing was observed during the transits of Comet 1882 II (when, however, the Sun was observed at a very low altitude) and Halley's Comet in 1910 (when a solid body of 50 kilometers would have been required to show on the solar disk—a size considered too large for a reasonable estimate of the nucleus of this comet). Nevertheless any such event, particularly in cases where the comet is close to Earth, should be carefully investigated (though certainly *not* by the novice), as should the possible—though to my knowledge never actually observed—transit of a comet across the disk of a planet, in the remote hope of observing a tiny black dot drifting across the face of the Sun or planet.

Secondly, a number of comets have been known to pass extremely close to the Sun, recent examples being Wilson-Hubbard (1961 V), Seki-Lines (1962 III), Pereyra (1963 V), Ikeya-Seki (1965 VIII), Daido-Fujikawa (1970 I), and White-Ortiz-Bolelli (1970 VI). Particles the size of average meteoroids would have completely vaporized at the least solar distance of these comets. In fact, the aforementioned comets of 1963, 1965, and Comet 1970 VI went so close to the Sun that objects the size of minor boulders may not have survived, yet the comets were not destroyed by their fiery dash through the heart of the solar system—although a certain *disruption* of the 1965 comet was noted at perihelion (i.e., the point closest to the Sun) and another one several days later.

Thirdly (and surely the most spectacular demonstration of all), it is now widely believed that either a small but still active comet or the remnant of a defunct one struck the Earth in 1908—the famous Tunguska fireball about which so much nonsense has been written lately.

On the morning of June 30, 1908, a brilliant fireball flew over Siberia accompanied by tremendous blasts which seemed to shake the very earth.[3] Moving from a southeasterly direction, the visitor trailed a thick column of dust which appeared, from points along the projection of the fireball's trajectory, to take the form of a gigantic pillar. Suddenly, over the remote Tungus region, the giant bolide exploded in a great ball of fire and smoke and a series of deafening detonations which could be heard over 1,000 kilometers away. Many kilometers from the center of the blast, houses shook, windows broke, and people were swept from their feet by the shock wave. Some 60 kilometers away, at the Vanavara factory, one witness was thrown several meters through the air, experienced a sensation of heat, and lost consciousness.

Seismic and barometric waves following the blast were registered on instruments as far away as England several hours later.

More peculiar, however, were the bright nights and strange sky effects observed over wide areas following the fall. For instance, the night immediately following the object's arrival was said to have been so bright that people living in the Caucasus found it possible to read newsprint out of doors and without artificial light at any time of the night. This effect persisted for about two months, slowly waning all the while.

Similarly, extraordinarily low levels of atmospheric transparency were noted during the same period, as were unusual sunsets and high cirrus clouds—evidently a lot of dust had entered into the Earth's atmosphere.

Strangely, despite the magnitude of the explosion and obviously large size of the meteoroid, no fragments were ever found and no crater marks the place of impact. What were originally believed to be small meteorite craters were later found to be of terrestrial origin and quite common in the region.

Meteoric dust, however, was found in the region and the scene of destruction gave a clue to the absence of cratering. The destruction had come from above—from a downward blast of air and an accompanying fire storm. Obviously, the bolide had exploded in flight without ever reaching the ground; a strange situation indeed for a massive meteoroid!

Clearly, the object was unusually fragile, and it is this fact which provides the strongest evidence for its cometary nature. Anything as dense as an asteroid should have left something more than dust; there would have to have been meteorites!

Also, the bright nights could suggest a cometary tail or alternately, if the comet had ceased to be active, a very rapid crumbling of the body during its flight through the atmosphere.

It has been argued that the approach of the bolide is suggestive of an object moving in retrograde orbit (this term will become clearer in Chapter 2. For now, it will be sufficient to note that the term refers to an orbit inclined at more than 90 degrees to the plane of the ecliptic and that such orbits are described by about half the known long-period comets and a very few medium-period comets, but by *no known* asteroid or meteorite—or, for that matter, by no comet of a very short period). If the Tungus object really was moving in such an orbit, this must be seen as very strong evidence in favor of its cometary nature. It would seem to follow immediately that the object was a small comet of the long-period (or just possibly, of the medium-period) variety.

Nevertheless, recent research by Dr. L. Kresak and a re-examination of the evidence available concerning the fall itself have cast doubt upon this interpretation and at the same time have raised another equally interesting possibility.[4]

Kresak contends that the orbit of the bolide was actually that of a short-period object rather close to aphelion (the point farthest from the Sun). Such an orbit would be direct and would resemble the Apollo asteroids (a small group of objects with perihelia inside Earth's orbit) or Encke's Comet.

Now, the radiant (the apparent point of origin on the celestial sphere) of the fireball, according to Kresak's findings, is very close to that of the β-Taurid meteor stream—a daylight meteor shower discovered by radar techniques and (like the nocturnal Taurids seen in November) associated with debris from Encke's Comet. Moreover, the β-Taurid stream peaks about June 30—the date of the Tungus event!

Thus there appears to be good reason for associating the Tungus object with Encke's Comet (which also passed perihelion, i.e, the point of its orbit closest to the Sun, in 1908). Kresak argues that the object was a fragment of the main comet which broke away many revolutions ago. In 1908, it would have passed perihelion about May 16 (some 15 days after Encke's Comet itself), at that time being within the orbit of Mercury. Like Encke's Comet, its aphelion lay beyond the orbit of Mars, and hence the object would have traversed a considerable portion of its orbit by the time it collided with the Earth.

Kresak estimates that the object which caused the Tungus event would have been about 100 meters in diameter and doubts if this is sufficiently large for cometary activity to have been apparent. In other words, he tends to the view that the object was a defunct, rather than an active, secondary comet. This would explain the lack of observations before the collision (although the object would have maintained a fairly small elongation from the Sun) and the absence of any previous discovery.

Nevertheless, recent determinations of the diameters of the nuclei of short-period comets give values in the general range of the Tungus object. Comet Pons-Winnecke and Comet Schwassmann-Wachmann 3 each have nuclear diameters of not more than 400 meters, and Comet Tuttle-Giacobini-Kresak may be as small as 200 meters. Moreover, these are not the intrinsically faintest (and, presumably, not the smallest) comets known—although they are, of course, at the fainter end of the scale. Midgets like Wilson-Harrington (1949 III) or the satellite of Biela's Comet may have been actually smaller than the Tungus object and yet revealed cometary activity. A dusty coma and tail could explain the odd sky illumination which followed the Tungus event.

The nondiscovery of the object need not be fatal either, as a faint comet having an orbit similar to Encke's would be at its brightest when close to the Sun in the twilight sky and may well have passed unnoticed.

Fourthly, it is a well-known fact that comets spawn millions of tiny particles which continue to orbit the Sun in the path of the parent body. When the Earth crosses one of these "dusty" comet orbits, many of these particles collide with the upper atmosphere and are destroyed in quick bursts of light. These are the meteors or "shooting stars" which can be seen on any clear night.

From time to time, very rich showers of meteors appear, the most famous and remarkable being the Leonids associated with the periodic Comet Tempel-Tuttle. This is an annual shower, but every 33 years when the parent comet returns, it becomes especially intense. The showers of 1799, 1833, and 1866 were particularly notable, while those of 1899 and 1933 were weaker—although one witness to the 1933 shower described it as appearing like a child's "sparkler" (a hand-held firework emitting silver dartlike sparks) held in the sky, presumably implying that the meteors were very frequent, very bright, quick, and silver.

The comet was observed in 1965 for the first time since 1866, and the following year the greatest meteor shower ever recorded burst upon the central-western U.S.A. For some twenty minutes, at the height of the display, meteors were appearing at the fantastic rate of *40 per second*—some of which were as bright as the quarter Moon![5]

At first sight, it may seem strange that no meteorites have been definitely associated with such showers (excepting the somewhat special case of 1908 as this, if indeed Kresak is correct, was a case of the collision of Earth with a fragment of the comet itself —a "secondary nucleus"—rather than with a meteoroid). Meteorites which have fallen at the same time as other showers (as, for instance, on April 4, 1095) seem to be only coincidental.

However, the conclusion drawn by R. A. Lyttleton and others, namely that comets *entirely consist* of very small particles, does not follow. Icy conglomerates would be very prone to break up in the atmosphere, even if relatively large, and would quickly disintegrate even if they did reach the ground.

By examining the behavior of meteors, astronomers can derive much information concerning density, tensile strength, and the like; and the behavior of shower meteors strongly suggests that they are just the weakly cohesive, low density structures that one would expect to originate from the icy conglomerate of Whipple's model.

Moreover, the majority of sporadic (nonshower) meteors reveal similar structures—even for fireballs of extreme brilliance—which together with some reliable determinations of cometlike orbits for these objects, strongly suggests that even meteors of considerable size originate in comets. Indeed, the most brilliant meteor observed by the fireball monitoring network was estimated to have had a mass of over 200 tons but was completely consumed during its flight through the atmosphere.[6] This fragility, not primarily the size of cometary meteoroids, explains the lack of cometary meteorites.

Then again, the oft-quoted cliché that large meteors and meteorites never fall from meteor showers is not strictly correct. Very bright fireballs do occur in showers (for instance, a Leonid fireball in 1966 was clearly visible in bright twilight and must have been caused by a large body), and there is a good chance that some of these may reach the Earth as meteorites. Apart from the occasional falls of blocks of ice which have been suggested as possible candidates for cometary meteorites (but mostly seem to have fallen off aircraft), recent research on fireballs occurring in the Taurid meteor shower shows that some of these have characteristics typical of meteorite-dropping fireballs—i.e., they have a brightness about as great as the full Moon, and the fireball appears to burn out before the meteoroid itself is totally consumed.[7]

It will be very interesting if one of these Taurid meteorites can be observed with sufficient accuracy to facilitate its recovery, as we would then almost certainly have a piece of Encke's Comet to analyze in our laboratories!

Some authorities, far from believing that no meteorites are of cometary origin, have maintained that a whole class—type 1 carbonaceous chondrites—may be of cometary origin, although many astronomers seem to be having second thoughts about this.[8]

Recovery of a meteorite of this class, or something of an even more fragile nature, would be an exciting find and would certainly answer, one way or the other, the vexing question of whether some carbonaceous chondrites are of cometary origin.

What a cometary meteorite is like is, then, an unsettled question; but whether it is a carbonaceous chondrite or not, the evidence is mounting that such a thing really exists and will find its way into a laboratory sooner or later, much to the delight of cometary astronomers.*

Fifthly, we may also note the strange behavior of Bradfield's Comet (1974 III).

Initially, this comet showed a strong dust coma and was easily observed in infrared light; however, between March 12 and April 5, the infrared emission suddenly and unexpectedly fell by some three magnitudes. About the same time, the appearance of the comet altered; the old central condensation was replaced by a sharp stellar nucleus. These results seem to indicate a rather sudden falling off of the comet's dust production. Initially, the comet was producing a large quantity of fine dust that was rapidly heated by the Sun's rays, which in turn accounted for the rather high infrared flux. It also helped obscure the central regions of the nucleus. However, this fine dust emission apparently gave way to the evolution of much larger particles which took longer to

* Perhaps one already has. Some rocks brought back from the Moon by Apollo 16 astronauts were high in volatiles and may be remnants of a comet which hit the Moon long ago.

heat up and hence accounted for the initial drop in the infrared flux. Soon even these particles stopped being produced, and the coma "cleared" of dust, allowing the nuclear regions to be observed for the first time.[9]

The picture of the nucleus suggested by this sequence of behavior is that of a layered structure—an outer crust containing a good deal of fine dust overlaying a layer of much coarser particles and finally a purely icy layer or else a layer of rock from which gases are evolved by desorption upon solar heating.

There is also the possibility of observing signs of the rotation of the nucleus, a possibility which, if confirmed, would surely indicate the presence of a solid body.

Observations of recurring jets and the like certainly imply rotation, but the most convincing evidence would be direct observation of periodic brightness variations of the *nucleus* (not variations in the comet's total light). The most suitable comets in which to search for this possibility would be those in which the nucleus is relatively bright in comparison with the total light—i.e., those objects which have a distinct nucleus but only a weak coma and little central condensation. Several objects of short period fulfill this requirement, but these are also comets with low intrinsic magnitudes and are, consequently, very faint objects only able to be studied with great difficulty.

Nevertheless, in 1976 one such object (the short-period d'Arrest's Comet) came unusually close to the Earth and actually became sufficiently bright to be observed with the naked eye. Moreover, it was very well placed for observation and could be followed throughout the night. This rare opportunity was seized upon by Fay and Wisniewski, who carried out accurate photometric observations of the nucleus and discovered clear evidence of brightness variations having periods of 5.17 ± 0.01 hours which they interpreted as being due to the rotation of an approximately pear-shaped solid body about 0.5–1.5 km in diameter. These observations were such that the contaminating light of the comet's gaseous coma was carefully eliminated, and the magnitude variations observed were unlikely to have been due to fluctuations within this feature. Moreover, other observers monitoring the comet's total light (principally the light of the coma) throughout this period did not record any variation in magnitude.[10]

Furthermore, the color of the nucleus was noted to be the same as that of the Sun, once again indicating a solid body reflecting sunlight and not merely a line-of-sight effect within the (in this instance gaseous) coma.

The picture of the nucleus as a solid icy body surrounded by a cloud of meteoric particles seems, therefore, to be the one most readily acceptable in view of the observational evidence. The older model of a cloud of separate particles and the "neo-classical" model of Lyttleton, in which a comet is seen as an extended cloud of very small particles (bread-crumb size or smaller) with no definite nucleus, cannot, in my opinion, account for the observed behavior of comets. (It may be relevant to remember that the "classical" model of a comet as a cloud of particles was initially put forward before powerful telescopes, capable of readily discerning the nucleus, were employed in comet study, and that Lyttleton, father of the neo-classical "sandbank model," admits to never having observed a comet through a telescope.)

The specific problems of the Lyttleton theory—which is basically a theory of the origin and dynamics of comets and only secondarily a model of their physical structure—will be taken up in the following chapter.

In addition to these rather specific issues, the model of Whipple alone appears adequate to account for the most general feature of comets, namely their ability to produce large amounts of gas time and again as they near the Sun. Explanations such as the desorption of gases occluded in meteorites (the classical) or evaporated during the collision of particles (Lyttleton's) simply do not allow a comet to go on producing gas at the rate observations show to be actual. Only an icy nucleus, it would seem, enables a comet to grow an adequate coma.

THE COMA

So far, we have been dealing with a part of comets which is, at best, inconspicuous and at worst, invisible. We were, nevertheless, justified in considering the nucleus first as it contains within its small volume most of the mass of the entire comet.

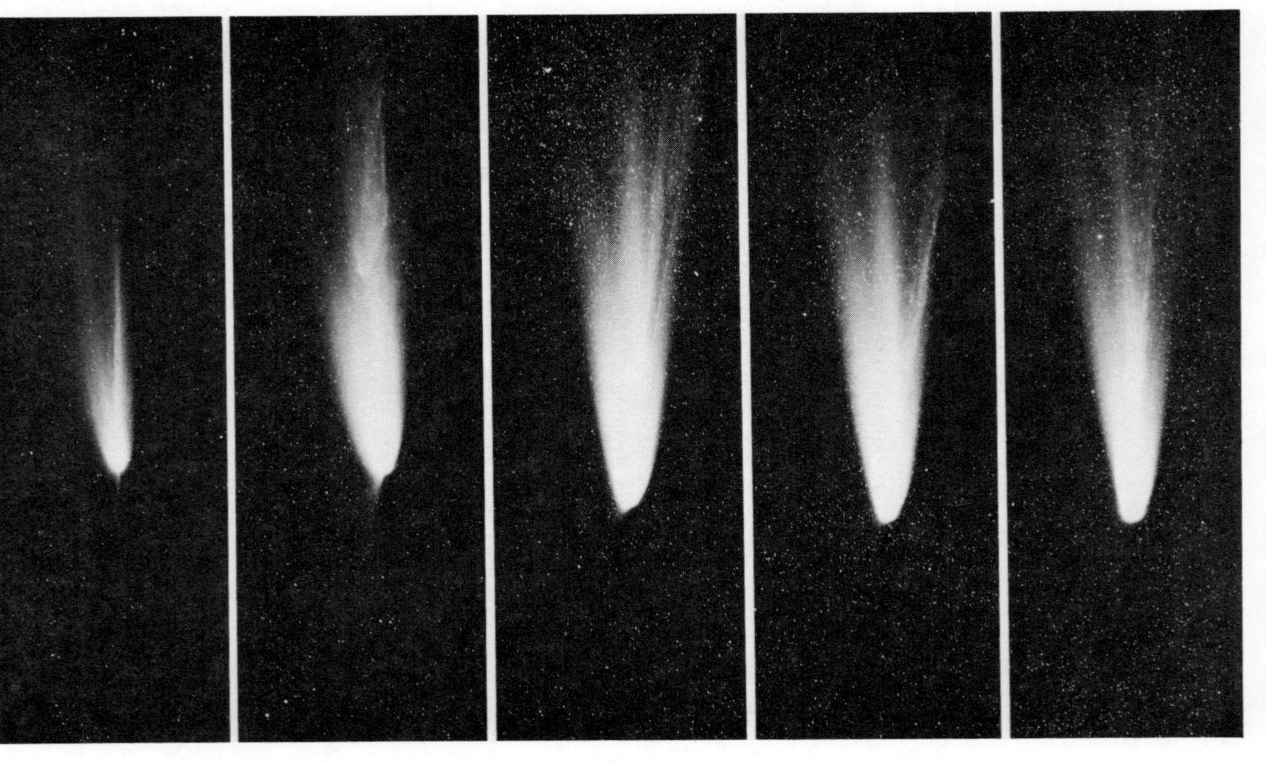

Plate 1. *Comet Arend-Roland (1957). Five views taken with 48-inch Schmidt telescope (April 26, 27, 29, 30, and May 1).*

As we move outward from the nucleus, we first encounter a somewhat larger cloud of particles, really an extension of the nuclear cloud, which frequently takes on an almost stellar appearance in small telescopes and may be (in fact, not infrequently is) mistaken for the true nucleus. This is usually simply referred to as the "central condensation" or, where it is especially well defined, the "false nucleus." The bright comets Arend-Roland of 1957 (Plate 1) and West of 1976 displayed very pronounced central condensations.

Beyond the nucleus and central condensation and extending thousands of miles into space is a vast cloud of incredibly low density, thinning out gradually at increasing distances from the nucleus until it eventually merges imperceptibly with the vacuum of space.

Spectroscopy reveals that unlike the nucleus and central condensation, the overwhelming proportion of light from the coma is due (in the vast majority of cases) not to reflection but to the excitation of gas molecules by solar radiation.

Thus the coma may be termed self-luminous, although its luminosity in the final analysis still depends upon radiation from the Sun. Electromagnetic radiation of one wavelength is absorbed and re-emitted in a process known as resonance fluorescence, similar to the excitation of gases in neon lights, the aurora, and galactic emission nebulae.

Using the Whipple hypothesis, the coma is quite readily explained. Thus an icy conglomerate moving toward the sun at first receives very little heat, until somewhere beyond 3–4 A.U. (c. 500 million kilometers), the more volatile substances start boiling away, surrounding the conglomerate with a nimbus of gas. As the comet draws ever closer to the Sun, less volatile material will start to appear in gaseous form and the more volatile will boil away at an ever-increasing rate.

Other processes will be at work as well. Radiation (chiefly ultraviolet) from the Sun causes some molecules to break down, forming new molecules and radicals (many of which could not exist for long in environments having greater gas densities) which can then be detected by terrestrial spectroscopes.

Moreover, as the ices evaporate, varying quantities of gas and stones are freed from the conglomerate, and these provide the dusty constituent of the coma. At very small heliocentric distances, these dust particles and meteoric stones themselves become gaseous, and the spectrum of the comet takes on the appearance of that of a meteor with several metallic elements seen in emission.

Such a picture is confirmed by spectroscopic analysis of the changes in the constitution of the cometary coma. At very large distances from the Sun, we generally find only the solar reflection continuum—i.e., the comet appears to consist only of solid particles. Presumably, some gas must have evaporated to release these particles from the frozen conglomerate, but the only candidate for such evaporation at distances of more than about 4 A.U. is methane, and this is difficult to detect in such low concentrations. Also it seems likely that the only comets to be observed at these distances are those with more than the usual share of very volatile substances—a suggestion which may help explain a not infrequently observed "failure" of those comets which are observed well before perihelion to come up to expectations as they near the Sun.

Generally the first gas to be identified is cyanogen (CN) with C_2 (carbon), the radicals CH and NH, and the amine group NH_2 appearing as the comet nears the Sun. Atomic hydrogen, carbon, and oxygen have also been detected and, at longer wavelengths, HCN, CH_3CN (hydrogen cyanide and methyl cyanide) and (very importantly, with reference to the Whipple model) H_2O. At close approaches to the Sun, sodium, calcium, chromium, cobalt, manganese, iron, nickel, copper, and vanadium have been observed. In addition, infrared observations have detected the presence of silicon in the continuous spectrum.

It will be apparent, therefore, that since the light of the coma is not due to simple reflection, its rate of increase with the comet's approach to the Sun—and its decrease with the comet's recession—will vary differently than that of the nucleus. Prediction of a comet's brightness is, therefore, no simple matter. It does not vary as, for instance, a minor planet shining by reflected sunlight, nor does it always vary in accordance with a constant formula; it is not unusual to find that a formula which gave fairly accurate predictions while the comet was far from the Sun, breaks down at smaller heliocentric distances—notable examples of this being

Plate 2. *Comet Cunningham, December 21, 1940. Photographed with 5-inch aperture Ross lens.*

Cunningham's Comet (1941 I) (Plate 2), Kohoutek's Comet (1973 XII), and Meier's Comet (1978 XXI), all of which gave initial promise of being much more brilliant objects than they in fact turned out to be.

We shall take up the question of brightness predictions later, when we come to deal with the general question of the brightness of comets. For the present, it is sufficient to note that the coma and nucleus are both physically and photometrically distinct. Figure 3 represents this in graphical form—it shows how the central condensation of Comet Arend-Roland and the nucleus of Comet Ikeya-Seki varied with respect to the total light of these comets.

Observations indicate that the nucleus of a faint comet far from the Sun contributes a greater portion of the total light than that of a bright and active comet at a smaller heliocentric distance. Put in a different way, it would seem that the light from a faint and

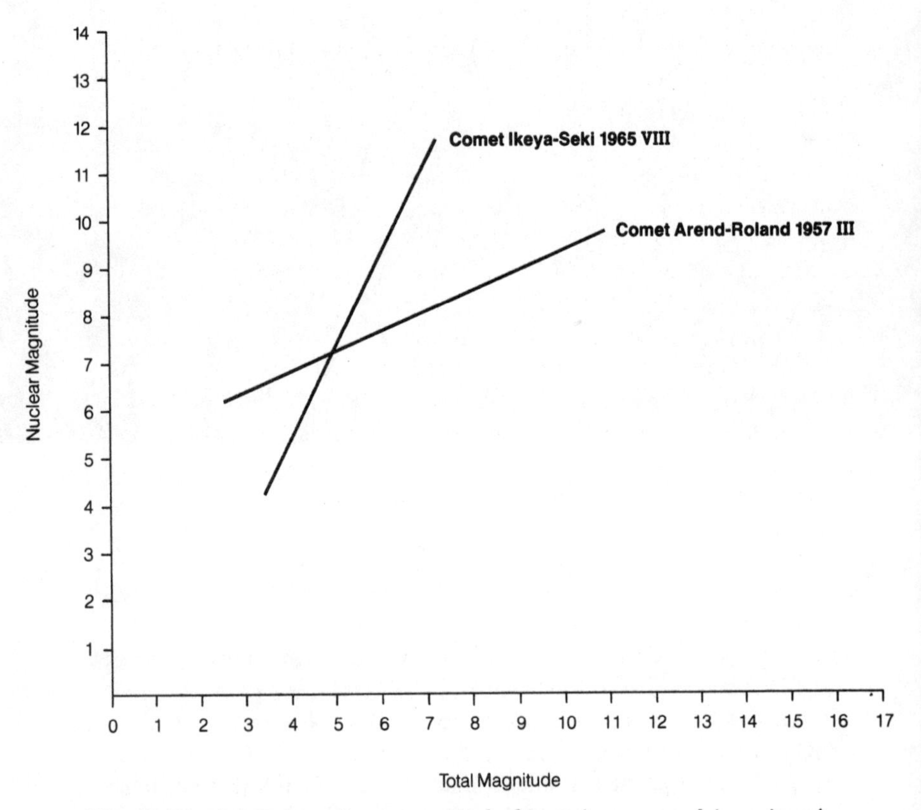

Fig. 3. *The brightness of astronomical objects is measured in units of magnitude, the units used on this graph.*

distant comet derives from a different source than that of a bright one. The comet as seen from a great distance is, in a very real and vital sense, *not the same thing* as a comet seen near the Sun!

Size of the Coma

We have seen that, astronomically speaking, the nucleus of a comet is a very minor thing. Most are less than 15 kilometers in diameter (3 or 4 kilometers is probably a fairly representative size for fairly large comets and less than 1 or 2 for small ones), and most would be less massive than a mountain on Earth.

In contrast to these very modest dimensions, the coma *is* an astronomically significant object, in size if not in mass, although (as with all features of comets!) there is considerable range between individual such objects and even in any particular object at different parts of its orbit.

Extremes in size range from virtually no coma at all in the case of Comet Wilson-Harrington (1949 III), one of the smallest and intrinsically faintest known, to some 2.2 million kilometers (far greater than the Sun) for Holmes's Comet (1892 III). This last was, admittedly, the result of a temporary eruption, but Flaugergues' Comet (1811 I) had a coma of some 1.7 million kilometers—still in excess of the volume of the Sun—and Ensor's Comet (1926 III) reached about 1.6 million kilometers at one stage, although this disintegrating comet had become very faint and of low surface brightness by the time such diameters were reached. More normal coma diameters lie between these extremes, averaging around 100 or 200 thousand kilometers—quite impressive objects indeed.

Comets passing very close to the Sun typically show a very small coma. For instance, Thome's Comet (1887 I) was mostly without any form of head or visible nucleus, although a small diffuse spot was observed on one or two occasions. It nevertheless displayed a long and impressive tail, as did Gould's Comet (1880 I) and Comet 1843 I, neither of which showed any appreciable coma when near perihelion. Similar behavior was also shown by Cruls's Comet (1882 II) and Ikeya-Seki (1965 VIII)—both of which displayed brilliant nuclei—and by White-Ortiz-Bolelli (1970 VI) and Pereyra's Comet (1963 V). Perhaps the strong "solar wind" at the very small heliocentric distances of these comets at perihelion, plus the very strong disruptive effects of sunlight upon gas molecules, prevented the formation of a typical coma.

Whether coma diameters have a general tendency to shrink as a comet nears the Sun is, however, an unsettled question. Certainly, as we have said, very small heliocentric distances tend to produce very small coma diameters, but careful measurements of diameters at distances greater than these has not led to any unambiguous conclusion. For instance, investigations by Wurm of Encke's Comet suggested a contraction with decreasing distance, whereas a study of this same comet by Vsekhsvyatsky failed to confirm Wurm's conclusion. Moreover, the behavior of Faye's Comet seems to suggest the opposite.[11] (See Appendix 1 for research suggestions and formulae for calculating the real dimensions of a cometary coma.)

The difficulty in estimating the size of a comet's head is, of course, the fact that (unlike the disk of a planet) there is no definite boundary between the luminous coma and the background sky. Surface brightness simply decreases in intensity until it becomes equal to that of the background sky. At this point we fix the boundary of the coma simply as a concession to our eyesight, and the size of this visible patch of light is assumed to represent the true dimensions of the coma. Photographs frequently extend the coma farther than the human eye, and even a change to a lower-power eyepiece with a larger field (and, consequently, greater concentration of light in the image) usually results in an increase in the size of the visible coma. It is another of the eccentricities of comets that they often appear larger in low-power eyepieces than in high-power ones!

Accurate measurements have shown very faint emission well beyond the boundary of the visible coma, revealing the presence of a much larger, far more rarefied, cloud of gas than that seen through a telescope or recorded on the photographic plate, and of course no earthbound instrument (irrespective of how accurate it may be) can trace the outer limits of the coma if these far-flung reaches of the comet radiate in an area of the spectrum to which our atmosphere is opaque.

Since the advent of observation satellites, we now know that the true size of the coma is vast compared with the (itself not inconsiderable) visual coma which we see from Earth. Comets, it has been discovered by means of extraterrestrial observation, contain considerable quantities of hydrogen, and it is this gas (radiating in the far ultraviolet regions to which our atmosphere is opaque) that has revealed the true size of the coma. Comet Tago-Sato-Kosaka (1969 IX), the first to be observed from outer space, revealed a hydrogen coma of some 1.6 million kilometers in diameter or larger, whereas Bennett's Comet (1970 II), for which confirmatory observations were made, displayed a hydrogen coma at least 12.8 million kilometers in diameter—at least 10 times the diameter of the Sun!

Initially, some astronomers speculated that these hydrogen comae may have been formed by the comet's "gathering up" the solar wind (itself a stream of hydrogen ions) at great distances from the Sun. Such a suggestion seemed possible (though perhaps a little far-fetched) as both 1969 IX and 1970 II were objects of long period (especially the former), but it would be severely tested if a short-period object could be observed from beyond the Earth's atmosphere. In fact, the short-period Encke's Comet (the last object one would expect to "gather" a cloud of solar wind particles) was observed from a man-made satellite during its 1970 return and was found to have a hydrogen coma. Clearly, the hydrogen must arise within the comet itself, and the suggestion that it comes from the photodissociation of the water molecule is in accord both with the Whipple model and with the subsequent discovery of water in Bradfield's Comet (1974 III) and of H_2O^+ (ionized water) in the tails of Ikeya's Comet (1963 I)—although it was not recognized at the time—and Kohoutek's Comet (1973 XII).

With the discovery of the hydrogen coma, the estimates of the gas content of comets must be increased about 100 times—adding, incidentally, further support for the Whipple model and confirming this model's estimate of the amount of gas produced by a comet based upon observations of departures from gravitational motion ("nongravitational effects") and the assumption that such departures were caused by the thrust of escaping gases.

It seems almost incongruous that such huge systems, in size comparable with large stars, should be associated with astronomically tiny objects like comet nuclei. However, the shock of this discovery is only a foretaste of what is in store when the tails of comets are measured, as there we face objects exceeded in size only by the very largest stars, nebulae, and whole stellar systems.

Appearance of the Coma

Anyone who has observed a number of comets cannot help but notice the great lack of uniformity in the appearance of these objects. A planet always looks like a planet; a star like a star; even galaxies and nebulae have a certain homogeneity of appearance which, while not as rigid as that of stars and planets, at least is sufficient to prevent confusion.

True, not all stars and planets look alike; nevertheless, Mars is not likely to be confused with Venus, and a planet like Jupiter could hardly be mistaken for a nebula like the Cygnus Loop! But the difference in appearance between, say, Jupiter and the Cygnus Loop is hardly greater than that between the mistlike smudge that was Kojima's Comet (1973 II) and the brilliant apparition known as Comet Seki-Lines (1962 III), brighter than Jupiter and, in appearance, just as solid.

In view of this, it may appear rather ambitious to attempt classification of the shapes of comets, but if a sufficiently large sample of objects is examined, various broad types do emerge which, to a certain extent, tend to coincide with differences in composition. These broad types are represented in Figure 4.

Two even wider forms can be detected in this (already very generalized) classification. The first form can be seen in (1), (2), and (3), examples of comets with globular comae (becoming pear-shaped in 3).

When this globular shape is found, especially in comets with small perihelion distances, spectroscopic analysis generally reveals strong gas emission and a relatively weak continuum, except in the immediate vicinity of the nucleus. The light of such comets is, therefore, contributed mainly by glowing gases with only a minor contribution from reflected sunlight.

In other words, the comet is a gassy one—a feature reflected in the rather weak, straight, but frequently very long and structurally complex tails produced by these comets. As will be seen later,

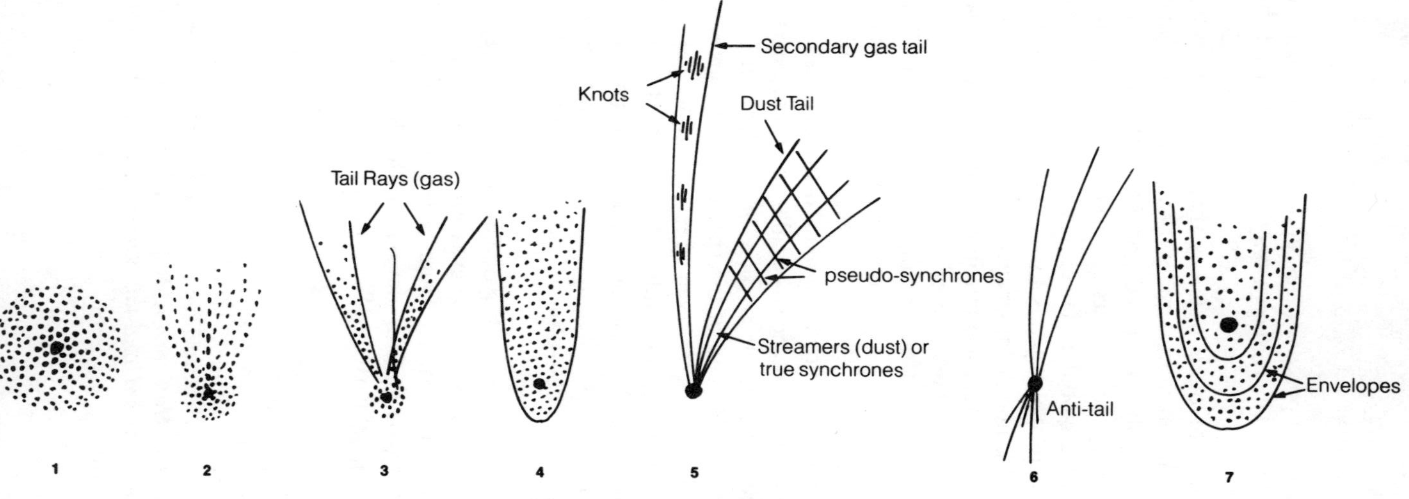

Fig. 4. *Comet classification.*

such tails are composed of gas and are quite dissimilar to the broad curving dust tails shown only weakly, if at all, by globular comets.

Another feature often noted is that these comets tend to increase in brightness more rapidly than usual as they approach the Sun and, conversely, fade rapidly as they withdraw from it.

Telescopically, such comets are usually rather featureless, looking rather like a globular star cluster, often containing a brighter (but hazy) core which tends to be more or less centrally placed in the coma (although fan-shaped comae, with the condensation at the apex of the fan, are not infrequent—a feature shown conspicuously for a short period by my own comet discovery, 1978 XV). Nevertheless, photographs (especially if made with blue-sensitive plates) very often reveal intricate and interesting tail features, seldom seen in the brighter dust tails, and wide-angle telescopes or binoculars show many of these comets to be beautiful and delicate objects—wraiths of low surface brightness.

Incidentally, the major portion of the light emitted by these comets is in the blue end of the spectrum and, consequently, they tend to appear bluish or even green in color. As globular star clusters are composed of reddish population II stars (i.e., older, metal-deficient stars), these objects tend to have a distinct *pink* appearance, and this difference in color is worth remembering when comet-sweeping—although it must never be taken as a foolproof hint!

Examples of comets with globular comae are Ikeya (1963 I), Ikeya (1964 VIII), Suzuki-Saigusa-Mori (1975 X), and the short-period Encke's Comet. Kohoutek's Comet (1973 XII), a dust comet prior to perihelion, developed the features of a typical globular gas comet early in 1974 (Plate 3).

Plate 3. *Comet Kohoutek (1973 XII), January 23, 1974. Photograph by Dr. H. Giclas, 33-cm. f/5 photographic refractor exposure time 30 minutes. Note the ribbonlike gas tail, the diffuse dust tail and the faint sunward anti-tail. Official Lowell Observatory photograph.*

Dust comets, particularly those of small perihelion distance or considerable activity, usually develop quite a different shape (Figure 4, nos. 4, 5, and 7). Here the head and tail merge imperceptibly, forming as it were a parabolic envelope with the nucleus at or near the focus. The most spectacular and brightest comets are generally of this form or some close approximation to it.

These parabolic heads are frequently scenes of strong activity, including jets of material erupting from the nucleus, parabolic envelopes in the coma, and relatively sharp parabolic outlines which contrast sharply with the indefinite boundaries of the globular objects.

The brilliant Bennett's Comet (1970 II), shown on Plate 4, is a fine specimen of this form of comet. In addition to a display of jets and parabolic envelopes, it featured a peculiar "Catherine-wheel" display of orange colored filaments in the near nuclear region. This has been interpreted by some astronomers as evidence that the nucleus was rotating. The "orange pinwheel," as it was described at the time, was observed both visually and photographically and reveals, in part, the extent of activity within the central regions of a truly active comet.

Dust comets and gas comets can be considered as the two extremes of a continuum. Most comets lie somewhere in between, having a coma composed partly of gas and partly of dust, each operating according to its own dynamic and each contributing its share toward the total light of the comet. The visual and photographic appearance of the comet will, of course, depend upon the relevant contribution made by each.

As most of the light emitted by cometary gases is confined to the blue region of the spectrum, filters transmitting only red light will reveal the extent and shape of the dust coma in light uncon-

Plate 4. *Comet Bennett (1970 II), March 21, 1970, 0240 hrs. U.T. (Universal Time), as photographed by M. J. Bester with 35.6-cm f/1.7 Schmidt telescope at Boyden Observatory, South Africa. Exposure time 10 minutes on Perutz film. Note the contrast between the curving dust tail and gas streamers.*

taminated by gaseous emission. Similarly, a narrow wavelength filter covering the region including the prominent emission feature known as the Swan Bands of C_2 (say 4,700–5,500 angstroms) will reveal, essentially, the gas coma with only a minimal contribution from sunlight reflected at that wavelength.

Photographs of Comet Tago-Sato-Kosaka taken with the aid of a filter transmitting the green of the Swan Bands revealed a bright, almost tailless spherical patch, whereas a photograph taken with the aid of a red filter showed a typical example of a dust comet with a sharply outlined parabolic head and an almost parallel-sided tail (see Plates 3, 4, and 5). This comet was a rather gassy one, although it displayed both gas and dust tails (superimposed, because of the geometrical line-of-sight effect, one upon the other) when close to the Sun. At greater heliocen-

Plate 5. *Comet Tago-Sato-Kosaka (1969 IX), January 10, 1970. This 5-minute exposure with the University of Michigan's Curtis Schmidt at Cerro Tololo shows the type I tail in the radiation of the blue-violet bands of CO^+ (ionized carbon monoxide). IIa-0 emulsion used. Note the bright central region of the tail, which is about 1½ million kilometers long on this plate.*

Plate 6. *Comet Tago-Sato-Kosaka (1969 IX), December 29, 1969. 103a-F emulsion through Schott RG1 filter, 3-minute exposure with the same instrument as for Plate 3. In this plate, the radiations of CO^+ (ionized carbon monoxide) and C_2 are blocked by the filter and only sunlight reflected by dust is recorded. Note the shape of the coma and the type II dust tail, the narrow dark lane at the center of the tail (sometimes incorrectly termed the "shadow of the nucleus"), and contrast these features with those shown in Plates 3 and 5. Nearly two million kilometers of tail are recorded on this photograph.*

Plate 7. *Comet Tago-Sato-Kosaka, January 11, 1970. This photograph was obtained with the Curtis Schmidt using 103a-J emulsion and a green interference filter showing the coma in the light of the green band of the C_2 molecule. Exposure time 5 minutes. The diameter of the coma recorded here is some 104,000 kilometers but only a trace of tail is recorded (extending up at roughly "ten o'clock").*

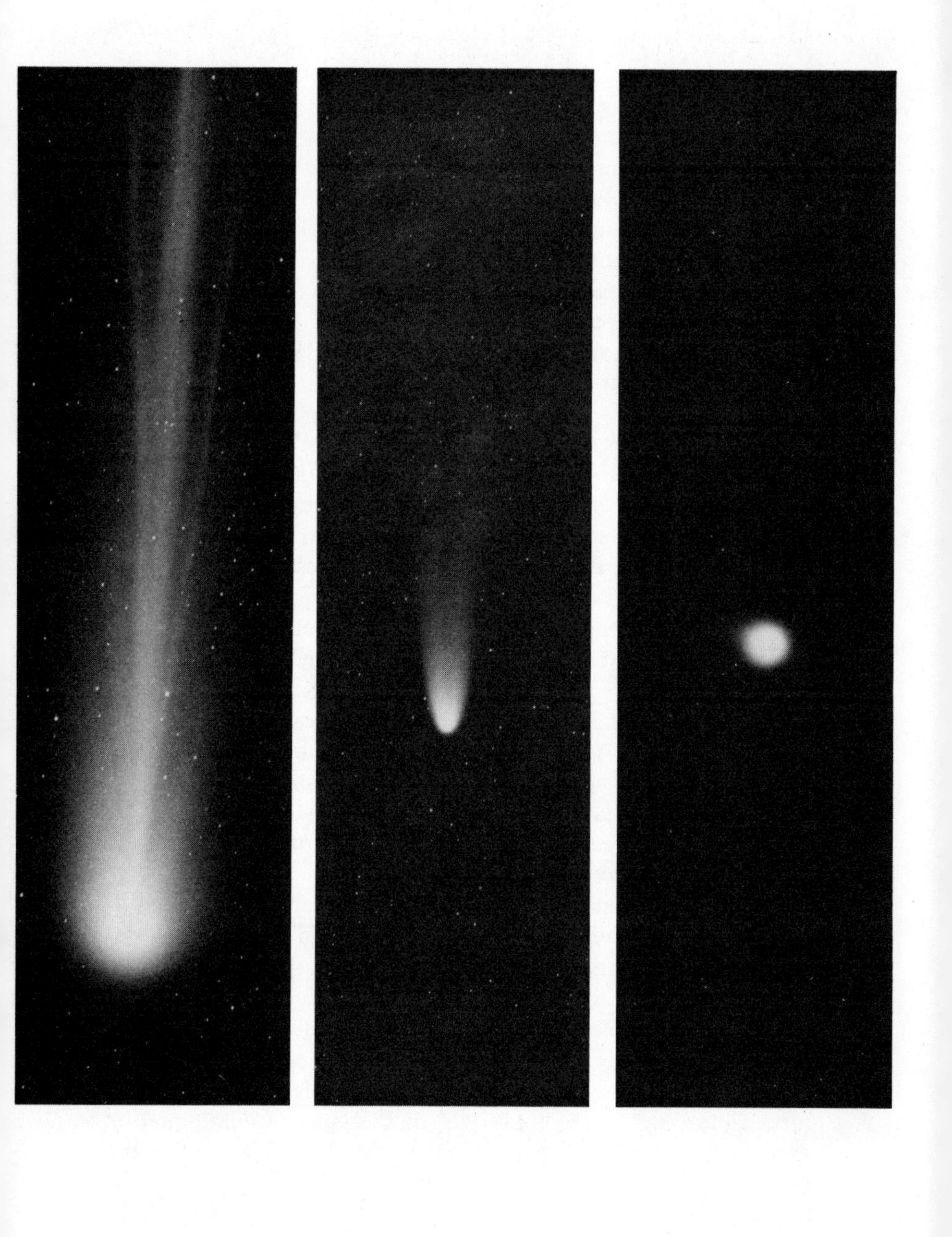

tric distances, it appeared as a fairly typical globular object with a faint, though complex, tail.

It must not be imagined that comets necessarily maintain their form throughout an apparition or that their shape is always as regular as might seem to be implied by this simplified account. Encke's Comet, for instance, frequently displays a fan-shaped coma with the "central" condensation displaced toward the sunward apex of the fan, and a teardrop shape (especially conspicuous on photographs) is not uncommon, as was displayed by both Ikeya-Seki (1965 VIII) and Kohoutek (1973 XII), in both instances before perihelion.

Brightness of the Coma

The magnitude of the coma is, except for those very faint comets which show only a nuclear condensation and little or no real head, the actual brightness of the comet as estimated by visual observers.

Many people seem to harbor wild ideas about the brightness of comets, believing that great comets like Halley's turn night into day (this expression was actually used in one children's encyclopedia article about Halley's Comet) and appear in our skies as enormous balls of fire. Actually, very few comets reach a brilliance in any way approaching that of the planet Venus and, of those which do attain this magnitude, most (for reasons which will soon be apparent) are so close to the Sun that observation is difficult, if not impossible.

The most brilliant of all comets—the major "Sun-grazers"—have been clearly visible in full daylight within two or three degrees of the Sun and, at this stage, have obviously been of fantastic brilliance—but then, they would need to be of fantastic brilliance to be visible at all under such conditions! If it were possible for a comet such as Cruls's or Ikeya-Seki to be visible in all its perihelial glory against a dark sky, the night truly would be turned into day, as the total light of the comet would equal or exceed that of the full Moon but would shine from an almost point-source nucleus with an intensity hundreds of times that of moonlight. However, as this is not possible, there is no sense in discussing it any further—the comets we see in a dark night sky are (except for a very few freaks which may have rivaled Venus) seldom as bright as the brightest stars. No comet has ever been definitely recorded as having cast a shadow—a feat easily performed by the star Sirius.

So the occasional story one hears about Halley's Comet "lighting up the night" should be taken with less than a grain of salt. It is a pity that this misconception has found its way into the minds of otherwise quite well-informed people, as it can only cause disappointment when Halley's next returns, in 1986. Actually, as we shall see in a later chapter, predictions concerning the 1986 return of Halley's are not optimistic about this object's reaching a very high brightness, with most estimates making the comet as faint as magnitude three or four at maximum, much less bright than a few dozen stars.

As we have already mentioned, the light of a comet comes from several different sources, all ultimately dependent upon the Sun, though not necessarily dependent in the sense of simple reflection in the manner of planets, satellites, and minor planets.

Furthermore, the coma is an evolving system, not a static one like a planet. That is to say, its constituents are continually being replaced; it is changing in size, in composition, and in density, and each of these changes must have some effect upon its brightness as seen from the Earth.

A minor planet, for instance, is a stable object. It remains (for all practical purposes) unchanged for thousands of millions of years. Its nature will not change in any way as it nears or recedes from the Sun, nor does the extra heat to which it is exposed near perihelion cause any temporary change in its intrinsic properties. It simply grows brighter in our telescopes as it nears the Sun and fainter as it recedes, corresponding to the amount of solar light it receives.

Once the absolute magnitude is known (i.e., the magnitude the planet would reach at $r = \Delta = 1$ AU, where r is the object's distance from the Sun and Δ its distance from Earth, both given in astronomical units), it becomes a simple matter to derive the apparent magnitude (i.e., the magnitude the object has for an observer on Earth) at any particular instant—assuming, of course, that the object's orbit is known. Shining by reflected sunlight only, its intrinsic brightness will vary according to the inverse square law. Furthermore, the distance from Earth will also modify its apparent magnitude, and this will also depend upon the inverse square law.

Thus, for planets the brightness H can be represented by the formula

$$H = \frac{H_0}{\Delta^2 r^2} \phi(\alpha)$$

where H_0 refers to the standard distances $\Delta = 1 \text{ AU} = r$, with Δ and r being distances from Earth and Sun respectively. $\phi(\alpha)$ is the appropriate phase-law function.

Expressed in terms of the usual stellar magnitude scale, this takes the form

$$m = m_0 + 5 \log \Delta + 5 \log r + \int(\alpha)$$

where m and m_0 are apparent and absolute magnitude respectively, and $\int(\alpha)$ is the modified form that the phase law takes.

Unfortunately for those wishing to make predictions about their magnitudes, comets are not so well behaved. Except for a few very faint objects which show little more than a nucleus, this above formula does not work for them. The changes in density and composition of the coma are not sufficiently regular to be predicted with any high degree of confidence, and furthermore, we are not sure just which processes are photometrically the most important ones. Such predictions are little more than trial and error.

In practice, though, it has been found that a simple interpolation formula gives fair approximations most of the time—though exactly *why* it does is not really obvious.

This formula is usually written in the form

$$m = m_0 + 5 \log \Delta + 2.5n \log r$$

where m_0 (absolute magnitude) and n are unknowns for any particular comet.

Notice that in the instance of a minor planet, the value of n is already known by the theory of the behavior of the brightness of objects shining by reflected light, and although not *explicitly* included in the earlier formula, is equal to 2 ($5 = 2 \times 2.5$). Upon this value, the further value of m_0 was calculated, and the formula employing these parameters could be used at all times to predict the magnitude of the minor planet.

For comets, however, the values of n differ not merely between objects but even for a single comet over different parts of its orbit. Moreover, it is seldom easy to determine the value of n given only a small set of magnitude estimates. Accurate estimation of the total magnitude of a comet is not an easy matter, and even among experienced observers estimates have been known to differ by as much as six magnitudes! Then again, even if we did manage to have an accurate set of estimates, the difference between light curves drawn according to divergent values of n may still not be apparent—the observations may fit a number of possible values of n. This was especially apparent in the instance of Kohoutek's Comet (1973 XII), which at a great distance from the Sun could have had its magnitude fitted by a number of curves of different n-values, including one having n equal to about 8 and the corresponding m_0 to about -2 and another having n around $+4$ and m_0 about $+5$. At a distance of nearly 4 A.U., either formula could be used to predict the approximate brightness of the comet; but at the heliocentric distance of the comet's near-perihelial section of orbit, the discrepancy was so vast that the one prediction gave a brightness just within naked-eye visibility, whereas the other would have rendered the comet visible in daylight!

The parameters n and m_0 do not, therefore, refer to "fixed properties" of comets and must be treated as handy aids to magnitude determination rather than as expressions of some invariant physical property.

It will be apparent to the reader that the determination of these parameters, even approximately, for any given comet involves a good sample of observations spread over a considerable arc of the orbit. Quite obviously this will not be satisfactory for anyone

wishing to predict the brightness of a newly discovered comet, and it is for just such objects as these that we are anxious to know this quantity.

For this reason, the computer of magnitudes generally makes the assumption that the comet's magnitude will increase (or decrease, if the comet is receding) according to the average value of n found for all comets—most frequently assumed to be close to 4. Absolute magnitudes are then computed according to this assumption.

The formula—which in most cases is *fairly* satisfactory—then becomes

$$m = m_0 - 5 \log \Delta - 10 \log r$$

Frequently, comets of short period are more sensitive to heliocentric distance than those of longer period, and for these, the value of n is often assumed to be 6 (i.e., "$15 \log r$" in the formula) or even, on rare occasions where the comet is altering rapidly in brightness, equal to 8 (i.e., $20 \log r$).

Those comets which are seen to have higher values of n are frequently ones with a large gas content and relatively little dust, although this does not invariably follow. For instance, the very dusty Bennett's Comet (1970 II) had n equal to 5 whereas the gassy Bradfield's Comet (1978 VII) had n equal to only 2, increasing in brightness only as would be expected for an object reflecting sunlight—though clearly it was *not* merely reflecting sunlight (see Plate 8).

The formula is usually fairly trustworthy at "normal" heliocen-

Plate 8. *Comet Bradfield 1978*c, *March 8, 1978, as photographed by G. R. Martin with the Curtis Schmidt at Cerro Tololo Observatory, Chile. Exposure time 5 minutes on IIa-O emulsion. This plate reveals a typical gas tail structure of several narrow rays diverging from the central region of the coma. Over four million kilometers of tail are recorded on this exposure.* (Photo courtesy of Dr. Freeman D. Miller.)

tric distances, but it becomes somewhat less certain at very small distances, and extreme magnitudes predicted on the basis of this formula for comets of very small perihelion distances are always to be taken somewhat casually. As we have already stated, the light from a comet at very small heliocentric distances comes from a different source than that of a comet at the more normal distances over which the formula applies. Comets near the Sun shine not by the light of gases released from ices in the frozen conglomerate, but from the evaporation of solid dust particles and even small meteoric stones. Perhaps a dusty comet will be brighter at very small distances than a gassy one.

Also, at Sun-grazing distances, much of the solar surface would be beyond the horizon for an hypothetical observer at the comet's nucleus, and therefore most of the solar radiation would "miss" the comet; this effect must tend to keep the brilliance more moderate than would be "predicted" on the basis of the formula.

Very close to the Sun, a comet possesses very little head—it is of very small angular dimension—and this might also place an upper limit upon its brightness. Indeed, Sun-grazers may grow to be as intense as the limb, i.e., the apparent edge, of the Sun itself (the maximum possible for an object deriving its light from the Sun) but be of such small size as to appear like a star near the Sun's edge—almost as intense as the solar limb. It is possible for a comet, under these circumstances, to reach apparent magnitudes of between −15 and −18, although they could not be more than 0.5 degrees from the Sun's center and, therefore, difficult to observe. Nevertheless Cruls's Comet (1882 II) was seen only a cou-

Plate 9. *Comet Kobayashi-Berger-Milon (1975 IX), July 29, 1975 (2200–2230 hrs. U.T.), Kodak Tri-X 35-mm film, 400-mm f/5.6 telephoto lens at Kapteyn Observatory, Roden (using 147-mm refractor at 450× as guiding telescope).* (Photograph by E. P. Bus and G. Comello, courtesy of Dutch Comet Section.)

This is another example of a gas tail, consisting here of a single ray emerging from a globular coma. Compare and contrast with Plate 6.

ple of hours before perihelion and Ikeya-Seki (1965 VIII) was observed through the perihelion passage itself, when the comet missed the Sun by only 0.0032 A.U. (0.0077 A.U. from the *center* of the Sun). Descriptions of the comet's brightness at the time indicate values in the range of the maximum mentioned above.

One final point should also be mentioned regarding this formula and its application to comets at very small distances. As it stands, the formula predicts a brightness of $-\infty$ at $r = 0$ or $\Delta = 0$. This is not only impossible, it is meaningless; and the logical and mathematical difficulty is not alleviated by the fact that the distances r and Δ are measured from the *centers* of the Sun and Earth (they are *heliocentric* and *geocentric* distances) and that zero values are, consequently, physical impossibilities. The fact that this logical *reductio ad absurdum* exists is further proof of the purely empirical nature of the formula and further indication that it cannot be trusted under all conditions—useful though it might be for most of the time.

Nowadays comets are observed by both photographic and visual means, with the visual (especially, in the case of a very bright comet, the naked eye) yielding the highest magnitude estimates. Photographic estimates, on the other hand, tend to be much lower and are most often estimates of the central regions of the coma rather than the total or globular magnitude of the comet. Visual and photographic estimates may differ by as much as six magnitudes for the same object.

It is now customary to give two sets of magnitudes, m_1 or the total visual magnitude, and m_2 or the magnitude of the nucleus; the latter being much lower in bright comets and varying according to a smaller numerical value of n in the majority of cases. This second magnitude value is the one of interest to photographic observers, whereas the first indicates how bright the comet appears when observed visually with the aid of small, low-power instruments or, if the comet is a bright one, the naked eye.

Fluctuations in Brightness

Thus far, we have been discussing the development of a comet's brightness as if it were an orderly affair, with the comet brightening steadily as it moves toward the Sun and fading again as it recedes from it. In practice, this is very often quite far from the true position.

Figure 5 shows an instance of brightness variation, in this example, Comet Ikeya-Seki (1968 I), not to be confused with Ikeya-Seki (1965 VIII). The magnitude of this comet fell, after perihelion, in a manner not expected from an empirical formula based upon the object's pre-perihelion performance.

Such behavior is not an uncommon feature of comets. It was also displayed by the periodic comet Giacobini-Zinner in 1972 and by Kohoutek's Comet (1973 XII) in early 1974, following a very brief anomalous *increase* in brightness at perihelion. Other comets have increased steadily in brightness even after perihelion passage, as for instance the short-period Comet Grigg-Skjellerup during its 1977 apparition. In the case of Halley's Comet in 1910, the comet was systematically brighter after perihelion than before, and the fading was also considerably less rapid than the corresponding increase in brightness had been.

Then we encounter the small-scale fluctuations which are not infrequently noted in the light curves of those comets which have been observed with sufficient accuracy to enable their light curves to be confidently drawn.

Figure 6 shows a typical set of fluctuations in the light curve of Comet Jurlof-Achmarof-Hassel (1939 III). The light curve of this comet is a particularly interesting one in that it reveals with rare clarity a relationship between at least this type of cometary brightness fluctuation and solar activity (the bottom curve of Figure 6 being the sunspot numbers).

The effect of solar activity upon the brightness of comets also appears to be apparent in the tendency of these types of fluctuations to decrease when a comet reaches high heliocentric latitude. It has even been argued that the mean magnitude of a comet decreases with increasing heliocentric latitude and consequent decrease in exposure to the solar corpuscular streams which are so active at lower solar latitudes.[12]

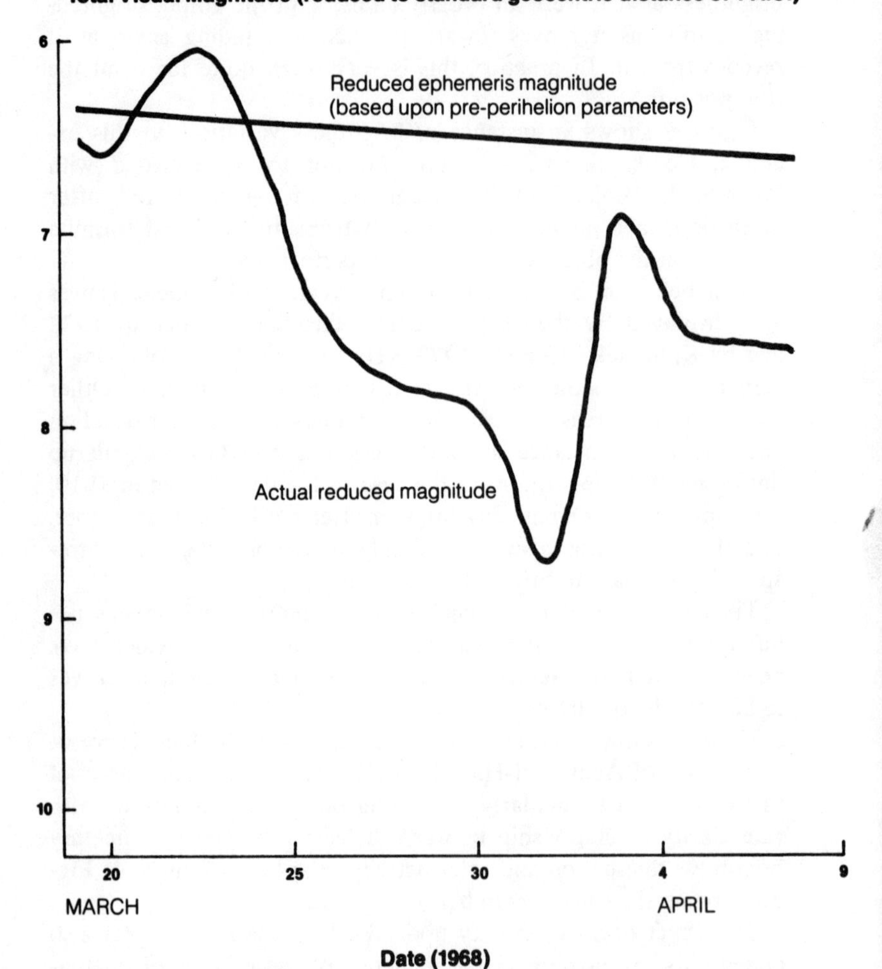

Fig. 5. *Magnitude variations of Comet Ikeya-Seki (1968 I).*

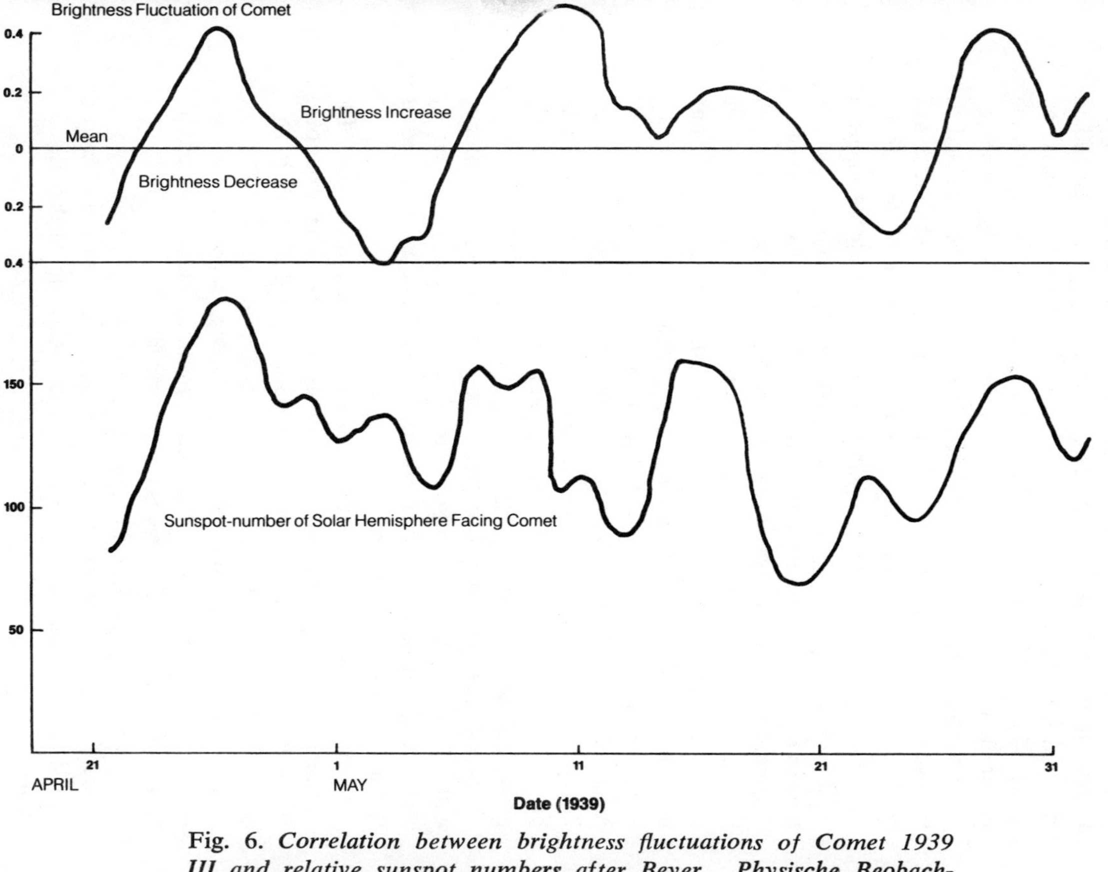

Fig. 6. *Correlation between brightness fluctuations of Comet 1939 III and relative sunspot numbers after Beyer, „Physische Beobachtungen von Kometen VIII“* Astr. Nachr. *282, p. 145.*

These types of variations are, however, very mild when compared with the behavior of some comets. For instance, Holmes's Comet during its discovery apparition of 1892 twice increased in size and brightness—on one occasion its brightness increased by some nine magnitudes, and its coma expanded to a volume greater than the Sun itself. Indeed, so mystifying did this behavior seem to the astronomers of the day, that Barnard even refused to believe that it was a comet at all, maintaining instead that it was really a dust cloud created by the collision of two minor planets. The absence of usual gaseous emission and the object's location in the asteroid belt supported this conclusion, which was nevertheless incorrect as subsequent returns of the comet have proven.

Periodic Comet Schwassmann-Wachmann 1 (which, although having a period of about 16 years, moves in an orbit of such low eccentricity that it may be observed annually) is especially prone to such outbursts and on one occasion increased its brightness by eight or nine magnitudes. Quite frequently it increases its brightness by up to five magnitudes, each "burst" following a similar pattern in which the nucleus suddenly becomes sharp and bright before swelling into a large coma and fading as it slowly grows diffuse.

Several attempts have been made to correlate these outbursts with sunspot activity, with some measure of success claimed by Richter, although many other astronomers remain unconvinced as to the reality of such alleged correlations.

Weird suggestions are made from time to time to account for the unexplained, and cometary outbursts have not been exempted from this practice. For instance, one line of speculation is that Schwassmann-Wachmann 1 is not a comet at all, but a minor planet composed of antimatter. The bursts, according to this theory, are really atomic explosions resulting from the annihilation on impact of meteoroids on its surface!

Until more conventional explanations fail or some strongly suggestive feature of the bursts is discovered (such as the powerful emission of gamma rays) this hypothesis needs no further discussion.

Another outburst-prone comet is Pons-Brooks, a fairly bright object with a period of about 71 years. Several outbursts were noted during the 1883–84 return, and in the less favorable

1953–54 appearance the comet again displayed wide and sudden variations from its mean magnitude, at times appearing brighter, at others fainter, than normal.

Other instances of similar behavior are provided by Humason's Comet (1962 VIII), in which an outburst of at least six magnitudes occurred when the comet was some 6 A.U. from the Sun, fully eighteen months after perihelion; Nagata's Comet (1931 III); and Comet Tago-Sato-Kosaka (1969 IX), which flared by about one or two magnitudes on February 7, 1970.

No talk of comet outbursts would be complete without some discussion of the great-grandfather of them all—the twin eruptions of the periodic comet Tuttle-Giacobini-Kresak during its return of 1973. This event was an important one as spectrograms of the comet taken during the second outburst revealed that most of the increase in light came from gases—not from an increase in dust as had been evident in both Holmes and Schwassmann-Wachmann 1. Evidently, there are "gas outbursts" and "dust outbursts," a fact which was already hinted at by the lack of alteration of the spectrum of Alcock's Comet (1963 III) during a much milder outburst in that year.

Tuttle-Giacobini-Kresak was first observed by Tuttle in 1858 as a small ninth-magnitude spot, but it faded before an accurate orbit could be calculated and was not seen again until 1907 when observed for a time by Giacobini as a dim object of the thirteenth magnitude.

The identity of these two comets was suspected, but nothing definite was established until Kresak discovered a 10.5-magnitude comet in 1951, which subsequently was found to be a return of the comets of 1858 and 1907. It moved around the sun in an elliptical orbit with a period of around 5.5 years, but its orbit was such as to render it unobservable at many returns.

It was seen again during the exceptionally favorable return of 1962, when it reached about magnitude 9, high in the sky for northern hemisphere observers.

The subsequent return was very unfavorable and no observations were possible. The 1973 return was somewhat more favorable, and the comet was again recovered on January 8 of that year.

Previously, the comet had given no indication of unusual activity and nothing very startling was expected in 1973. Although well placed in the sky, the comet was not particularly close to the Earth, being some three times more remote than it had been in 1962, and the magnitude was not expected to increase beyond about 13 or 14.

On May 20 the comet was of magnitude 14, visible only in fairly large telescopes or with the aid of long-exposure photographs. However, one week later it was visible to the naked eye as a fuzzy spot of about fourth magnitude; an increase in brightness of some 10,000 times in seven days! The magnitude was about 6.5 on May 20 and had fallen to 10.0 by June 1; and, according to John Bortle, it was 10.2 on June 3, when a short faint tail was visible. Thereafter the brightness fell off steadily, decreasing to 13.2 on June 23 and 14 or 15 by July 4.

On July 6, however, a new outburst had begun, increasing the magnitude to at least 6. It was during this second outburst that spectroscopic observations were made, revealing strong gaseous emission and a medium-to-weak continuum.

The comet had fallen to magnitude 9.8 by July 12 and 13, and the last visual observation seems to have been my own on July 21, when the comet appeared as a very diffuse patch of about magnitude 10.5. A photographic observation by D. Latham and C. Y. Shao on July 31 revealed that the magnitude had diminished to about 15.

This outburst was truly spectacular, but it is interesting to remember that the comet passed relatively far from the Earth in 1973. Had the eruption taken place in 1962, the magnitude would have increased to about —1, possibly rendering the comet visible in daylight with suitable equipment!

Clearly, the mechanism behind outbursts of this nature is something much more violent than that responsible for the comparatively mild fluctuations in the light curves of comets such as 1939 III.

Several theories have been put forward to account for these flares, the following being three relatively recent speculations:

1. Abnormal concentrations of very volatile material just below the surface of a cometary nucleus may sublimate to form

pockets of gas acting as centers of stress beneath a relatively inactive crust. The explosive release of these high-pressure pockets removes segments of the crust and exposes fresh volatile ices to the incident solar radiation. In those comets where the icy conglomerate has a high dust content, the outrush of gas may carry away such large quantities of dust that the resulting outburst will be largely due to sunlight reflected off dust, whereas other comets may experience a surge of gas and relatively little solid matter. This is in accord with observations, as we have already seen.

This explanation has been criticized from time to time, on the grounds that (assuming a nucleus of the Whipple type) outbursts should always accompany the splitting of the nucleus. However, I rather doubt that this is necessarily correct, as small segments of the nucleus could easily split away and crumble rapidly without becoming visible as a secondary nucleus. Conversely, there is no reason why an irregular nucleus should not evolve in such a way that irregularities may not simply "melt" off rather in the manner of an irregular piece of ice when exposed to a hot fire.

On the other hand, it is often noted that brightness flares and splitting of nuclei are associated. We have already seen that this was the case with Tago-Sato-Kosaka, and a similar phenomenon was noted accompanying the breakup of West's Comet (1976 III).

2. Explosive chemical reactions might occur, involving free radicals such as CH, OH, or NH. It is even possible that NH could convert to NH_4N_3 (ammonium azide) at a temperature of 148 degrees Kelvin (the temperature of a comet at about 2.5 A.U. from the Sun). This compound is violently explosive and, according to this line of reasoning, may trigger the outburst.

Another possibility is the formation of hydrogen peroxide (H_2O_2) from OH (the hydroxyl radical) at 77 degrees Kelvin. Diluted by stable compounds, these explosive chemicals are relatively harmless, but as evaporation continues their concentration increases until a point may be reached where some localized supply of energy (possibly a burst of radiation from the Sun) is all that it needed to trigger this cosmic "bomb."

One difficulty faced by this hypothesis is the lack of any observed flash immediately preceding an outburst, as would be required if such phenomena were really the result of chemical explosions.

3. Comets formed far from the Sun would, according to the theory of Harvey Patashnick, Donald Schuerman, and Georg Rupprecht,[13] contain not ordinary ice but a form known as "amorphous ice." At temperatures of less than 95 degrees Kelvin and pressures below $1N/m^2$ (a force of one newton per square meter), ordinary ice would, these authors argue, not form. Furthermore, amorphous ice suffers a phase transition to cubic (ordinary) ice at temperatures near 140 degrees Kelvin, releasing considerable energy in the process and abruptly increasing its volume. Amorphous ice is twice as dense as cubic ice, and such a transition must result in considerable stress if it occurs within an icy conglomerate nucleus.

The authors of this theory suggest that comets should show a tendency to suffer outbursts at a distance of around 2.5 A.U. from the Sun—such distances being in the region where temperatures of about 140 degrees Kelvin are reached. Some statistical evidence, they claim, supports this contention and, indeed, it may be relevant to note that comets as unalike as Halley and Arend-Roland (the first a periodical object, the second a "new" comet) have undergone outbursts at this section of their pre-perihelial orbits.

Nevertheless, amorphous ice has usually been formed in laboratory conditions of considerable purity, and it is not clear whether quantities sufficient to account for cometary outbursts are able to form under cometary conditions.

All things considered, I would give my vote to the first theory—or some close approximation to it—if I was asked to vote on the matter, at least until something better comes along.

Vanishing Comets

While cometary outbursts may be the most violent departures from the orderly magnitude development implied by the interpolation formula, some comets behave in a manner which, if less violent, is surely more drastic—they vanish altogether!

For instance, the great comet 1843 I, which when near perihelion was one of the most brilliant and spectacular ever seen, faded with unusual rapidity some weeks after perihelion, apparently disintegrating or, at least, exhausting its supply of volatile constituents.

Then we have the case of Alcock's Comet (1959 VI), nearly as bright as a star of the sixth magnitude at discovery and expected to become quite brilliant as it passed close to the Sun (it was even searched for in broad daylight on the day of perihelion —unsuccessfully, however). Not only did it not become brilliant, it failed completely to reappear in the twilight after perihelion and was only followed for one week after discovery. Similarly, Comet Daido-Fujikawa (1970 I) failed to reappear after passing close to the Sun (although the comet was rather badly placed), as did the Sun-grazing Comet Du Toit (1945 VII); and Winnecke's Comet (1874 I), also an object of small perihelion distance, was observed to become very diffuse, lose its point nucleus, and vanish as it neared the Sun.

Not only objects of small perihelion distance suffer this way, however. For instance, Comet Westphal faded out as it approached perihelion in 1913 and was not seen at its subsequent return in 1975; and the comets Ensor (1926 III) and Perrine (1897 III) also suffered similar fates, even though their perihelion distances were not very small.

Sometimes a comet is seen to split in two or (on rare occasions) into several components. Taylor's Comet (1916 I) split in 1916, for instance, and was for many years listed as lost—until one of the fragments was recovered in early 1977. Similarly, the short-period object Brooks 2, in 1889, threw off at least four satellite comets after passing through the satellite system of Jupiter three years earlier, but only the main comet survived the apparition. Evidently small comets formed through the breakup of larger ones do not normally persist for long, and it would seem that many comets splitting into two nearly equal parts must be approaching the end of their lives.

The most famous case of cometary breakup and disintegration, the one which really gave the phenomenon scientific respectability, involved the short-period Biela's Comet. In 1846, this comet was observed to have divided into two distinct segments, presumably having undergone schism some time since its previous return. At its subsequent apparition, both comets were again visible, first one and then the other being the brighter. That, however, was the last time the comet was definitely observed (although there have been a few, probably spurious, unconfirmed observations at various times since then), and it is generally believed that it completely exhausted sometime after the return of 1852.

The comet has nevertheless contributed to our list of celestial spectaculars by providing the great Andromedid meteor storms of 1872 and 1885. It is not strictly correct to say, as some popular works do, that these storms were the result of the breakup of the comet if by "the breakup" is meant the 1846 disruption. They seem (according to H. A. Newton) to have actually originated in a close approach to Jupiter in 1841 and may, therefore, have originated *simultaneously* with the breakup rather than as a further stage of disintegration. In fact, the storm might still have occurred if the comet had remained intact. Andromedids were observed prior to the comet's disruption—for instance in 1741, 1798 (when a strong shower was recorded by Brabdes), 1830, and 1838. In fact, old records suggest that the Andromedid radiant has been an active one as far back as 524 A.D. It should also be noted that strong meteor storms are also associated with comets which have not faded out, as for instance the Draconids of 1933 and 1946 (associated with the short period Comet Giacobini-Zinner) and the Leonids (with Comet Tempel-Tuttle).

Another famous instance of cometary disappearance, this time not involving visible disruption, was that of Pajdusakova's Comet (1954 II). Discovered as an eleventh-magnitude object on December 3, 1953, this comet was found to have a perihelion distance of only 0.072 A.U. on January 24, 1954. As we saw earlier, brightness generally increases rapidly as a comet nears the Sun, and an approach as close as this should have yielded a quite high brightness. The comet was expected to reach negative magnitude (i.e., to be brighter than zero magnitude) at perihelion, and although the head would probably be invisible, astronomers were advised to keep watch for the tail rising out of the dawn twilight.

Unfortunately, nothing of the sort happened. Throughout December, the comet increased in brightness at a rate not merely less than the average comet, but even below that of an object shining by reflected sunlight! Even worse, any slight increase in brightness which did occur was more than offset by the increasingly diffuse nature of the comet and the consequent diminution of the intensity of its light.

By January 8, 9, and 10, Van Biesbroeck and Jeffers failed to observe the comet at all, though it should have been observable even if as faint as magnitude 14.

The usual (and to my mind, most probable) explanation of this cometary phantom is simply that Pajdusakova's Comet was originally an intrinsically fairly bright comet which exhausted and probably disintegrated as it neared the Sun. However, an alternate suggestion has been proposed by N. B. Richter,[14] who suggested that this comet may have been very faint originally and was actually undergoing a violent outburst of light at the time of discovery. He further proposes that a significant proportion of so-called unconfirmed comet discoveries may refer to similar events. Nevertheless, the general behavior of Pajdusakova's Comet, compared with ordinary outbursts on the one hand and with cometary disintegration on the other, seems to me strongly to favor the more usual hypothesis.

Curiously, the twenty-first anniversary of this event was marked by another, closely similar, instance of the same phenomenon.

On November 13, 1974, John Bennett (an amateur noted for his discovery of the great comet of 1970) found a ninth-magnitude object moving south in the morning sky. Over the following few mornings it brightened by about one magnitude and became centrally condensed, with very slight traces of tail development starting to show.

Orbital calculations revealed the comet to be over two weeks from perihelion, when it would be between the orbits of Earth and Venus (in terms of distance). Some nine days after passing perihelion, it would pass closest to the Earth and was expected to attain a brightness of at least sixth magnitude. As it would then be rather near the south celestial pole (and consequently high against a dark sky for southern hemisphere observers), it would probably be visible with the naked eye if one knew exactly where to look.

Unfortunately, the comet had other ideas. As it neared perihelion, the coma grew increasingly diffuse, the central condensation weaker, and the intensity of light less. Total magnitude fell, but estimates made by different observers became increasingly at variance with each other as the coma grew more diffuse. Some astronomers estimated a fall of about five magnitudes from shortly after discovery until about November 25, but others estimated that the total brightness had remained almost constant, although the comet was increasingly harder to see due to the diffusing of the coma.

After November 25, the comet ceased to be observable, possibly due to moonlight at first, although it refused to show during a total lunar eclipse on November 29 and was not found in the dark skies of early December. A twenty-minute exposure by Dr. E. Roemer on December 16 likewise failed to show the comet, even though a *stellar* object of magnitude 18 could hardly have escaped notice.

There the position rested until two astronomers, C. Torres and J. Parra of Chile, found that they had recorded the comet's "ghost" on plates taken on December 8 and 10, about the time when it should have been brightest. All that remained was a very faint nebulous mass, recorded as a dim "stain" on the background sky, some 5 by 15 minutes of arc in extent and oriented northwest to southeast. Evidently the comet had simply melted away and diffused into space.

Rapid fading is not uncommon after a comet passes perihelion, and in certain instances this seems to involve the genuine exhaustion of the comet.

Such an instance was provided by Bradfield's Comet (1978 XVIII), an elusive object discovered on October 10 of that year, after perihelion. This object was approaching Earth at its discovery, and it was expected that the magnitude would remain a fairly constant 8 or 9 throughout the month as the comet's intrinsic fading was compensated by its decreasing geocentric distance. Nevertheless, this did not happen. The coma quickly became very diffuse, and the total magnitude had fallen to 10 or less some two weeks after discovery, according to Mr. David Herald of Canberra; later in the month it could not be found at all, even though it should have been well placed for southern hemisphere observers. Like Bennett's of 1974, Bradfield's Comet had simply faded away, never to be seen again.

THE TAIL

Not all comets show the feature which popular belief considers characteristic of these strange objects. Some appear simply as small luminous clouds. Of course, these may actually have tails that are too faint to be discerned or that radiate at a wavelength to which our eyes are not sensitive. Indeed, long-exposure photographs, especially those employing plates sensitive to blue light, frequently reveal tails on comets which visually appear devoid of such appendages.

On the other hand, certain comets have displayed tails which, even by astronomical standards, are impressive objects, as a glance at the following list of tail lengths will demonstrate.

1843 I	2.15 A.U.
1680	2.01
1769	1.50
1811 I	1.30
1965 VIII	1.30
1882 II	1.25
1910 II	1.14
1970 VI	0.86
1910 I	0.50
1861 II	0.34
1819 II	0.18

A comet tail usually has the visual appearance of a dim searchlight and, when clearly visible, is one of the finest astronomical spectacles. Any comet passing well within the Earth's orbit and having an absolute magnitude of about 8 or brighter should be watched as a good potential tail-grower, although (as with most features of comets) predictions are very hard to make—some comets passing near the Sun have little in the way of tail development, whereas others coming to perihelion at considerable distances sprout fine appendages.

In general, it is the presence or absence of a well-defined tail which determines whether a comet will or will not be an impressive sight—an object too faint to be seen with the naked eye may be a magnificent sight in binoculars while another, magnitudes brighter, may simply appear as an amorphous blob. For instance, Bradfield's Comet (1974 III) appeared, in binoculars, not unlike a great comet as seen with the naked eye, due to the magnificent bright tail which this relatively faint naked-eye comet developed near perihelion, in contrast to the much brighter Ikeya's Comet (1964 VIII) which appeared throughout as a large diffuse mass with a sharp central condensation but only a very faint tail.

Nevertheless, even a bright comet tail is less substantial than it appears. At least once in historic times (in 1861), the Earth has passed through the tail of a comet, and it probably encountered the edge of the tail of Halley's in 1910 and even, technically, that of Suzuki-Saigusa-Mori in 1975, yet no substantial effects were noted. (In fact, given the enormous size of the hydrogen coma of a large comet, it is probable that the Earth has passed through several of these as well during the course of recorded history. The daylight comet of 1472 is a possible instance, and yet we are still here to talk about it.)

Broadly speaking, comet tails are of two types: long straight bluish streamers composed of gas (type I tails) and broad curving dust tails which, unlike the first type, shine by reflected sunlight rather than by excitation of gas molecules and ions and are, therefore, more or less the color of sunlight. Dust tails are classified as type II or type III depending upon the degree of curvature, the latter being short appendages of extreme curvature. For most purposes, the difference between these two types of dust tail may be neglected and type III tails considered as extreme instances of type II tails.

A further type should also be recognized, namely anomalous tails or "beards"—i.e., appendages which are, or which appear to be, directed toward the Sun. These may be further subdivided into true anomalous tails (i.e., tails in which the matter is actually between the comet and the Sun, an example of which is provided by the tail of Ensor's Comet, 1926 III) and pseudo-anomalous tails (i.e., those in which the sunward appearance of the tail is a projection effect only, as for instance the famous "anti-tail" of Comet Arend-Roland, 1957 III). Sekanina demonstrated that the anti-tail of Kohoutek's Comet (1973 XII) was mostly due to projection.[15]

In addition to these types of tails, "late" and "distant" tails should also be mentioned as special variations, as we will see later.

Type I Tails

About 50 per cent of those comets which develop visible tails in the neighborhood of the Sun show a type I tail, sometimes in conjunction with a type II tail which may or may not (depending upon the projection considerations) be visually separate.

Type I tails were long a mystery for workers in the field of cometary forms. Whereas dust tails (as will be seen shortly) can be quite readily explained as resulting from the pressure of sunlight "blowing" tiny dust particles into a stream directed always away from the Sun, type I gas tails appeared to involve repulsive forces of an entirely different order of magnitude.

Thus, given in terms of units of solar gravity, the acceleration required to explain type I tails is of the order of 10^2 or higher, as against repulsive forces of approximately 1 in dust tails (i.e., the force of repulsion is 100 times greater than the gravitational attraction of the Sun, as experienced by the particles in these tails). Moreover, gas tails display activity—at times quite violent—and show firm evidence of variations in the repulsive force, which runs counter to what we would expect to find if repulsion were due to sunlight alone.

That the repulsive force originates in the Sun is, however, apparent by the fact that type I tails are most conspicuous, and most active, at distances smaller than about 1.5 A.U. Although they are sometimes observed at greater distances—as, for instance, in Humason's Comet (1962 VIII) and Morehouse's Comet (1908 III) at distances in excess of 2 A.U.—it is only much nearer to the Sun, generally within the orbit of the Earth, that full-fledged type I activity is usually revealed.

Such activity typically takes the form of long thin streamers or rays seemingly emerging from the outer extemities of the tail and converging toward the center, dying out upon reaching the central axis of the tail. Photographs sometimes show the central regions of the tail to be quite dark compared with the outer reaches (though at other times a *bright* central spine is noted).

Such features as these tend to give the tail a symmetrical appearance, rather like a series of concentric cones (as, presumably, what we see as a ray is really a cone projected against the two-dimensional background of the sky). However, such regularity is often destroyed by condensations, waves, and even, as in the remarkable case of Swift's Comet (1892 I) on April 7 of that year, by the appearance of "clouds" which seem to behave as separate comets even to the extent of possessing their own secondary system of rays. (This type of *tail formation* must not, incidentally, be confused with the appearance of real satellite comets like those of Brooks 2.)

Sometimes a cloud or condensation will accelerate to speeds far in excess of the usual repulsion, as notably in Morehouse's Comet (1908 III). At other times, whole sections of the tail will become detached and the comet will grow a new tail (as, for example, the comets of Morehouse and Humason), and when activity is very high, sharp edges develop which may persist for several days, as they did in Halley's Comet (1910 II) and Mrkos's Comet (1957 V).

Clearly, such events cannot be explained by the simple pressure of sunlight. Much more powerful forces are at work. Indeed, so peculiar and vigorous does this activity become that (in the past) suggestions have even been made ascribing cometary activity to some unknown substance. As is now known, this suggestion was not quite as wild as one might think. Not that comets are composed of unknown *material,* but comet tails—at least type I tails—can only be satisfactorily explained in terms of a *state* of matter unknown until fairly recent times and very rarely found on Earth. This is the "fourth state of matter"—plasma.

Having, of course, nothing whatever to do with either blood plasma or Victorian seances, plasma is a gaseous state of matter in which the atoms have been stripped of their electrons. Such a plasma constantly surrounds the Sun, especially at low heliocentric latitudes, in the form of the "solar wind"—a stream of protons forming, as it were, the outer fringes of the solar corona. This "wind" blows furiously at the Earth's distance from the Sun, but slows to a "breeze" at about 2 A.U. and, presumably to a "zephyr" at greater heliocentric distances, maybe even "piling up" as a cloud of ionized hydrogen at distances greater than Pluto—a suggestion put forward by Dr. W. I. Axford.

Now, the observational fact that fully developed gas tails are virtually always within the radius of the zone of transition from "solar wind" to "solar breeze," plus the further observation that these tails are less frequent at high heliocentric latitudes (where the "wind" blows more quietly) is strongly indicative of a connection between tails of this type and the plasma wind.

L. Biermann has, in fact, demonstrated the existence of such a connection, whereby the momentum of the tail gases is achieved by the transfer of momentum from the "solar wind" plasma.[16]

The tail itself can be regarded as a plasma. Spectroscopic analysis reveals the main constituents to be CH^+, CO^+, $CO_2{}^+$, $N_2{}^+$, OH^+, and H_2O^+ (ionized CH radical, carbon monoxide, carbon dioxide, nitrogen radical, hydroxyl radical, and water). Such a plasma may be expected to be rather sensitive to magnetic fields, and, indeed, we have direct evidence to this effect in the persistence of long filamentary rays. If these features were to form without a magnetic field to inhibit their diffusion, they would very rapidly spread to much greater widths than are observed in fact, due to the dispersing effect of the rather high temperatures which must exist within the tail plasma.

In addition to the normal "solar wind" and magnetic fields, gas tails also appear sensitive to powerful beams of corpuscular radiation emerging from active regions on the surface of the Sun. Indeed, whenever a suitable comet is exposed to these searchlights of radiation, enhanced tail activity almost invariably results, as it did in Swift's Comet (1899 I), Halley's Comet (1910 II), and Comet Whipple-Fedtke-Tevzadze (1943 I). In Halley's Comet, the largest observed accelerations occurred during times of pronounced "solar wind" activity, and at the same time there was an enhanced tendency for the formation of condensations in the tail. Likewise, in Comet 1943 I, high velocities were observed and strongly turbulent activity seen in the tail on two days, separated by an interval equal to the period of a single solar revolution. Similarly, for Swift's Comet, correlations were found between solar activity and the very considerable changes and brightness fluctuations experienced by this object. It is interesting to note that this comet appeared at a time of minimal solar activity, with only one active region on the Sun's surface at the time.

It seems, therefore, that type I tails are very sensitive solar wind-socks and for this reason, they have drawn the attention of solar as well as of cometary astronomers. In the days prior to artificial space probes, they were indeed the only means we had of observing solar corpuscular radiation in interplanetary space—nature's own space probes.

Type II and Type III Tails

As mentioned earlier, the difference between type II and type III tails is minor and can, for most purposes, be overlooked.

While not holding much interest for solar physicists or being as photographically interesting as type I tails, these "dust tails" are nevertheless more prone to provide better displays for the visual astronomer. This is because shining by reflected sunlight, they provide strong images in all regions of the spectrum, including those to which our eyes are sensitive, in contrast to gas tails whose major constituent, CO^+ (ionized carbon monoxide), radiates in the violet, a region of the spectrum to which our eyes are not particularly sensitive.

Furthermore, dust tails are not as dependent upon heliocentric distance as gas tails and are frequently observed in conjunction with comets passing perihelion at 2, 3, or 4 A.U. from the Sun. Thus, when Van den Bergh's Comet was discovered on November 12, 1974 (over two months after its perihelion), it showed a fairly pronounced tail 2 minutes of arc in length (fairly long in comparison to the coma diameter of only 8 seconds of arc), even though its *perihelion* distance was in excess of 6 A.U.!

Tails of such very distant slow-moving comets (let's call them "distant tails") are usually straight, with near-parallel and fairly well-defined sides forming a parabolic envelope with the head. They lack the curvature displayed by dust tails nearer the Sun, but are distinguishable from the straight type I tails by their lack of rays or other features and by the fact that they generally lie at an angle halfway between the radius vector (i.e., the line from the Sun to the comet) and the comet's orbit.

Recently Zdenek Sekanina has concluded that these distant tails are composed of ice grains of somewhat larger size than normal dust-tail particles, and may, therefore, be said to represent yet another type of comet tail, although the dynamic (if not the physical) considerations will be similar to those of dust tails in general.[17]

It should be remembered that many comets must pass at distances of 4, 5, or 6 A.U. from the Sun and that very few will ever be discovered due to their extreme distance. Nevertheless, when all these are taken into consideration, it might be true to say that these ice-dust tails are the most frequent of all comet tails.

Before leaving the subject of distant tails, a few words should be said about another variation—"late" tails. These, as their name suggests, develop well after the comet has passed perihelion and are seen at a relatively large heliocentric distance. The short-period Borrelly's Comet is especially prone to this sort of activity, and the late tail which it develops tends to show little regard for the radius vector. Thus, although it is traditional for a comet receding from the Sun to go away tail first, Borrelly's tends to go away head first, so that at times the tail may even point toward the Sun.

Sekanina's research on this phenomenon led him to conclude that such late tails are composed of large dust particles released near perihelion and appear to be associated with fan-shaped formations of the coma in the near-perihelial parts of the comet's orbit.[18]

The more conventionally observed dust tails typically appear as broad curved appendages increasingly lagging behind the extended radius vector, rather like the smoke plume of a speeding steam locomotive on a calm day. Where a gas tail is also present, the latter generally emerges as a straight ray from the side of the curving dust tail. Bennett's Comet (1970 II) displayed such a double tail. The line-of-sight effect was such that about two weeks prior to perihelion the comet was seen in the southeastern sky with a bright curving dust tail and a faint, straight, gas tail separating away from it at a distance of about 1 degree from the head. As the comet neared perihelion, the aspect changed until the dust tail to some extent eclipsed the fainter streamer. After perihelion, the gas tail emerged again into view as the dust tail continued to rotate north, and it was widely observed in early April.

Consider a comet nucleus of the Whipple model moving toward the Sun. Ices evaporate and boil off into the vacuum of space, freeing in the process formerly imprisoned solid particles ranging in size from rocks to dust specks to submicroscopic motes the size of the wavelength of light. These all leave the nucleus at the same low velocity (neglecting, for the moment, those blown out in a localized eruption), but their subsequent history will depend upon their size and density. As soon as they leave the protection of the frozen nucleus, they are exposed to two opposing forces:

1. Solar gravity trying to pull them toward the Sun.

2. Solar radiation pressure, attempting to force them away from the Sun.

Large particles succumb to (1), very small ones to (2). It is these latter which are swept away from the Sun and into the dust tail.

At any given moment, there will be particles of various sizes and densities leaving the nucleus, the smaller ones being driven farther by radiation pressure. The locus of (i.e., the path described by) these particles of all sizes is termed a *synchrone*.

If, on the other hand, we were to consider the locus of particles of a particular size and density that have been leaving the nucleus continuously, we would have another curve termed a *syndyne*.[19] For any moment of time, a family of syndynes can be drawn, each corresponding to a different value for the ratio of radiation pressure acceleration to solar gravity. Very minute particles of low density will describe syndynes which are closest to straight lines, whereas for larger particles, the syndyne will be more markedly curved. Syndynes of maximum and minimum curvature determine the edges of a type II tail, particles moving in paths of intermediate curvature "filling in," as it were, the space in between.

A family of syndynes drawn for still larger particles accounts for type III tail formation, and particles larger and heavier than these describe paths lagging increasingly behind the radius vector and approaching nearer and nearer to the comet's orbit until, finally, we have relatively large particles affected only negligibly by radiation pressure and consequently showing no movement in the direction of the extended radius vector. These particles continue in the comet's orbit under the influence of solar gravity. They are the meteoroids, those little bodies which burn out as shooting stars as they collide with our atmosphere. Meteor streams are, therefore, on a continuum with the dust tails of comets—one could even say that they represent one extreme curve followed by dust particles, the other extreme being the almost straight paths, closest to the extended radius vector, of very small particles. Most of the tail particles describe paths of intermediate curvature and lie somewhere between the extended radius vector and the path of the comet's motion.

Particles intermediate in size between normal dust-tail particles and the larger meteoroids, spread out in a thin fan which under certain conditions (namely, when the Earth crosses the plane of the comet's orbit after the comet has passed through perihelion) may be seen from Earth as an *anti-tail*—i.e., one which appears to point toward the Sun (to be more precise, a pseudo–anti-tail, as the apparent sunward direction is, in such tails, a projection effect rather than a true sunward extension of the tail). Anti-tails have been observed for the comets of 1823, 1882 II, 1957 III (Arend-Roland, the most famous and probably the finest case on record), 1969 IX, and 1973 XII—to name but a few of the better known. Comet 1974 III (Bradfield) even showed *two* anti-tails at one stage, though neither was very prominent. Wilmot's Comet (1844 III) is also generally cited as an instance of an anti-tail; however, Sekanina has argued that observations of this are inconsistent with synchrone and syndyne calculations and, apparently, refer to some other type of phenomenon (possibly some form of jet or eruptive event within the coma).

It is interesting to note that photographs of comets Arend-

Roland and Kohoutek (1973 XII), both of which provided fine examples of anti-tails, revealed the fan-shaped cloud of particles as a faint, diffuse glow in the sky on one side of the comet.

The particles responsible for these features in both comets were, according to synchrone/syndyne calculations, submillimeter or even millimeter in size. In the case of Kohoutek, infrared observations also showed the anti-tail particles to be of this order of size (they were far cooler than the particles in the dust tail and must, therefore, have been considerably larger) in apparent confirmation of these theoretical calculations.

It is interesting to remember that a normal "shooting star" is caused by a meteoroid the size of a grain of rice. Some of the particles in the Kohoutek anti-tail must have been sufficiently large to produce faint meteors had they struck the Earth's air, and it is therefore true to say that in the anti-tail of this and other comets, we actually witness the emergence of a meteor stream from a comet.

Very dusty comets are, of course, the ones most likely to produce major anti-tails, but such development is apparently not inevitable if we take into consideration the instance of Bennett's Comet (1970 II), which did not produce this feature when the Earth crossed the plane of its orbit after perihelion. On the other hand, gas comets are not exempt from anti-tail formation (though it cannot be expected to be spectacular), as is evidenced by Bradfield's Comet (1975 XI). This comet appeared first in the predawn sky well south of the Sun and increased to about magnitude 5 as it traveled into the twilight. However, the tail remained faint and short throughout this time. "Surprisingly little tail for a comet so bright and condensed" was the judgment I entered in my observing diary for December 9 and 11. My last observation (and, it would seem, the last observation before perihelion) was on the morning of December 12 when the comet looked almost like a slightly fuzzy, bluish star, deep in the twilight. It came to perihelion on December 21, and Japanese observations soon after perihelion found the comet to be between first and third magnitude when it began to emerge in the evening twilight as a northern hemisphere object. It faded quickly, but was fairly well observed and seemed to be about as bright as it had been in early December, by the time it was again generally visible in the evening sky. It was then that the faint anti-tail was noted as well as a faint normal tail, which appeared to have been a little more prominent than the pre-perihelion tail formation noted by myself. The impression I receive from the descriptions of the comet at this time is that the tail and anti-tail were approximately equal in prominence, although both were quite faint. At least, the tail does not appear to have been conspicuously brighter than the anti-tail.

Dust tails, as we have already remarked, are generally not as structured as gas tails. Nevertheless, as photographs of comets which were accompanied by considerable dust tails (for instance, West's and Bennett's) reveal, they are not necessarily without structure. Long diffuse streamers were very prominent in West's Comet, for instance; and it can be shown that these may be represented by synchrones of particles arising in "puffs" (presumable explosive or eruptive features) within the nucleus. These synchronous features are clearly distinguished from the gas streamers in that they are yellowish rather than bluish and are broader and more diffuse. Also, they are not given to the contortions which may affect gas streamers or rays.

Another, rare pattern noted in dust tails has had a history of controversy. I refer to those series of striations, looking somewhat like a series of ripples, which are oriented almost *across* the tail. These are approximately parallel and appear evenly spaced, the pattern covering on occasion the entire length of the tail. Nevertheless, the pattern always seems to be short-lived, persisting for a few days at most and seldom re-forming. Comets in which these features have been noted include the great comets of 1744, 1858, 1901 I, 1910 I, and 1965, as well as Mrkos's Comet (1957 V), Comet Seki-Lines (1962 III), and West's Comet (1976 VI)—all of which were very bright objects with intense dust tails.

Traditionally, such striations have been explained in the manner of the dust streamers—i.e., as synchrones. Nevertheless, this approach was strongly criticized by Vsekhsvyatsky, who argues that the mechanical theory fails to account for these features for three reasons:

1. The striations form a definite sequence of streaks or bands of approximately the same length and breadth, and are separated by approximately equal distances and have about equal intensity.

Such regularity is hard to explain in terms of random explosions in the nucleus.

2. The striations, if extended, converge at a point between the nucleus and the Sun, an observation not in accord with the traditional theory.

3. The striations and patterns of streaks are always short-lived, another observed feature not readily accounted for by the mechanical theory.

As an alternative, Vsekhsvyatsky[20] proposes that the striations are composed not of dust, but of very heavy polyatomic molecules which are then affected by magnetic fields and align themselves in the observed patterns. Unfortunately, spectroscopic analysis of dust tails is inconsistent with this suggestion, although (as pointed out by Wurm[21]) the suggestion that magnetic fields play a role in the formation of these false synchronous or pseudosynchronous features may hold some truth even if we believe (as we seem forced to believe on the weight of spectroscopic evidence) that the material in the striations is dust. Exposed to the powerful blast of solar radiation, the dust particles may be expected to lose electrons through the photoelectric effect and may, therefore, be influenced by magnetic and electric fields.

Nevertheless, it now seems probable that the mechanical theory need not be given up entirely and that speculation regarding magnetic fields may be unnecessary.

From observations of the active dust tail of West's Comet (1976 VI), Dr. Zdenek Sekanina[22] was able to show that the false synchrones formed an angle of some 10 degrees to the true synchrones, and in agreement with Vsekhsvyatsky, he concluded that the striations were indeed *false* synchrones. Nevertheless, these striations all originated in the diffuse edge of the tail which (through calculations of syndynes) could be shown to consist of relatively large particles emitted at perihelion; and, Sekanina reasons, it is probably *these* particles which are the immediate parents of those composing the striations.

That is to say, the striations are composed of dust derived from the breaking up of relatively large particles—in a sense, they are composed of the tails of myriads of microscopic comets. Confirmation of this thesis is provided by the relatively high brightness of the striations, indicative of large numbers of very small particles. In Comet Ikeya-Seki (1965 VIII), for instance, the portion of the tail where the striations were most conspicuous was also the brightest region, indicative of fine dust reflecting the Sun's light. This thesis looks to be the most promising yet postulated as an explanation for these curious features of comets.

REFERENCES

1. "Comet 1969g Observed Around the World." *Sky and Telescope*. Vol. 39 No. 4, April 1970, p. 265.

2. Zdenek Sekanina. "Relative Motions of Fragments of the Split Comets III. A Test of the Splitting of Comets with Suspected Multiple Nuclei." *Icarus*. Vol. 38 No. 2, May 1979, pp. 300–16.

3. Kirill P. Florensky. "Did a Comet Collide with the Earth in 1908?" *Sky and Telescope*. Vol. 26 No. 5, Nov. 1963, pp. 268–69.

4. *Sky and Telescope*. Vol. 56 No. 6, p. 497.

5. "Great Leonid Meteor Shower of 1966." *Sky and Telescope*. Vol. 32 No. 1, Jan. 1967, pp. 4–10.

6. Harold B. Ridley. "Meteorites." *Journal of the British Astronomical Association*. Vol. 89 No. 3, April 1979, pp. 219–38.

7. Keith B. Hindley. "Taurid Meteor Stream Fireballs." *Journal of the British Astronomical Association*. Vol. 82 No. 4, June 1972, pp. 287–99.

8. Harold B. Ridley. "Meteorites."

9. Edward P. Ney. "Multiband Photometry of Comets Kohoutek, Bennett, Bradfield, and Encke." *Icarus*. Vol. 23, 1974, pp. 551–60.

10. Theodore D. Fay and Wieslaw Wisniewski. "The Light Curve of the Nucleus of Comet d'Arrest." *Icarus*. Vol. 34 No. 1, April 1978, pp. 1–9.

11. Nikolaus B. Richter. *The Nature of Comets*. London: Methuen, 1963, p. 58.

12. Ibid., p. 137.

13. Harvey Patashnick, Donald Schuerman, and Georg Rupprecht. *New Scientist.* Aug. 15, 1974.

14. Nikolaus B. Richter. *The Nature of Comets,* pp. 142–43.

15. Zdenek Sekanina. "The Prediction of Anomalous Tails of Comets." *Sky and Telescope.* Vol. 47 No. 6, June 1974, pp. 374–77.

16. L. Biermann and R. H. Lüst. "Comets: Structure and Dynamics of Tails." In *The Moon, Meteorites, and Comets.* Edited by B. M. Middlehurst and G. P. Kuiper. Chicago: University of Chicago Press, 1963, pp. 618–38.

17. Zdenek Sekanina. "The Prediction of Anomalous Tails of Comets."

18. Zdenek Sekanina. "Surface Structure of a Cometary Nucleus." *Icarus.* Vol. 37, 1979, pp. 420–42.

19. Zdenek Sekanina. "The Prediction of Anomalous Tails of Comets."

20. S. K. Vsekhsvyatsky. "On the Nature of 'Synchronic' Formations in Cometary Tails." *Russian Astronomical Journal.* Vol. 39, p. 290.

21. K. Wurm. "The Physics of Comets." In *The Moon, Meteorites, and Comets.* Edited by B. M. Middlehurst and G. P. Kuiper. Chicago: University of Chicago Press, 1963, p. 600.

22. Zdenek Sekanina. "Disintegration Phenomena in Comet West." *Sky and Telescope.* Vol. 51 No. 6, June 1976, pp. 386–93.

2

Discoveries, Orbits, and Origin

Each year may bring a dozen or more comets into the scrutiny of Earth's telescopes, some being recorded for the first time (mainly those of very long period) and others being scheduled returns of known objects having shorter periods.

Some method of identification, therefore, becomes necessary. It is, of course, traditional to name a comet after its discoverer; however, while this is of indispensable value for periodic objects seen at several returns, it is not of such value for comets seen once only, especially in view of the fact that some comet discoverers seem to be quite addicted to finding new objects—M. Honda, for instance, has found twelve! Thus, to avoid confusion, a new comet (or a returning one that has just been recovered) is provisionally labeled with the year of discovery followed by a lowercase letter denoting the sequence within the year of discovery (or recovery)—"a" for the first comet, "b" for the second, and so on for that year. Thus Ikeya's Comet of 1963 was also known as Comet 1963a, being the first comet discovered that year.

Later, when a definitive orbit has been calculated (usually about two years later) using all available observations, this provisional listing is replaced by another, employing the year of *perihelion passage* (which might or might not be the year of discovery) followed by a roman numeral indicating the order in which the comet passed perihelion. Thus, Ikeya's Comet is now known as Comet 1963 I, being the first known comet to pass perihelion in 1963.

On two occasions (comets 1961 X and 1963 IX) the numerals do not follow the order of perihelion, as these two objects were actually found several years after perihelion on old photographic plates—after final listings for their respective years had been made.

As already mentioned, comets are named after their discoverers. In cases where there is more than one discoverer, the names of the first three are bracketed together, leading sometimes to such awkward appellations as Bakharev-Macfarlane-Krienke or Peltier-Schwassmann-Wachmann. Where the name of one discoverer is itself hyphenated, it seems that only the first part of the double name is used. Thus the object discovered by Balley-Urban and Clayton became known, simply, as Comet Balley-Clayton. Similarly, when two discoverers have the same name, the commonsense solution of using the name only once applies, as in Cesco's Comet, jointly discovered by C. U. and M. R. Cesco.

Sometimes a comet is named for other than the discoverer, as for instance the periodic objects Halley, Encke, Lexell, and Crommelin, named after the mathematicians who first established the short-period nature of their orbits.

One strong departure from the normal naming procedure was the adoption of the names "Tsuchinshan 1 and 2" for two periodic comets discovered in 1965 and "Tsuchinshan" for the nonperiodic 1977q. "Tsuchinshan" means "purple mountain," the name of the Chinese observatory where these discoveries were made.

New comets are discovered both by means of deliberate visual searches and accidentally on photographs taken for other purposes. During the 1950s photographic discoveries were in the lead, whereas the sixties saw a great upsurge in visual finds, mainly by amateurs (especially in Japan) sweeping the sky with small low-powered telescopes and binoculars. Early in the seventies, the pendulum swung back again in favor of photographic discoveries, with the new southern hemisphere observatories taking their fair share. Indeed, S. W. Milbourn of the British Astronomical Society saw this as a rather ominous tiding for the amateur astronomer; nonetheless, I still dared to be optimistic about the situation, noting that of all the comets discovered photographically since 1970, only two (1973 XII and 1976 VI) became sufficiently bright to be readily observed by visual means, whereas all other bright objects were discovered by amateurs. The tilting of the scales in favor of photographic discoveries was simply due to the paucity of bright objects during the first half of the decade, and once the bright objects started appearing again, the amateurs came back onto the scene as lively as ever—as is evidenced by the spectacular performance of William A. Bradfield of South Australia who has been the sole discoverer of comets 1972 III, 1974 III, 1975 V, 1975 XI, 1976 IV, 1976 V, 1978c, 1978o, 1979c, and 1979*l*.

Comet-sweeping is, therefore, still a potentially useful pastime which may be enjoyed by anyone with access to a low-power telescope or a pair of large binoculars, a clear dark sky, and plenty of patience, or who simply likes the relaxation of sweeping the skies.

It is difficult to specify the most suitable instrument for comet-sweeping; personal preferences rule supreme in one's choice and one's advice. Sufficient to say that as a matter of fact, most visual discoveries have been of objects of the tenth magnitude or brighter, irrespective of the size of the instrument used. Thus, an instrument which makes a tenth-magnitude diffuse object sufficiently conspicuous to be noticed during casual sweeping (for your eyesight and at your locality) would be suitable for comet-sweeping. In practice, a field of 2°–3.5° and a firm tripod are also highly advisable.

My own personal preference is for large binoculars, my own pairs being 20 × 65 mm and, more recently, 15 × 80 mm, to which cardboard dew caps and a black cloth hood (to pull over the observer's head—a necessity in my streetlight-infested locality) have been fitted. A pair of toy binoculars (acting as a finder) were fitted for a while, but were found subsequently to be superfluous. Mounted on a firm tripod—actually a machine-gun mount, now converted to more peaceful purposes—these binoculars will show comets to about magnitude 11 under good conditions.

In my opinion, the ability to use both eyes while observing is more restful than squinting through one eyepiece and the fields of 3° and 3.5° (for the 20 × 65 mm and 15 × 80 mm respectively) are adequate without being too large. The apertures are not large but seem to be adequate, and binoculars tend to accentuate faint diffuse nebulosities.

Thus far, I have discovered one comet with the aid of the 15 × 80 mm (Seargent's 1978m) and independently found three others (whose positions—and even existence—were unknown to me at the time) with the 20 × 65 mm.

One feature of a good comet-hunting instrument is often overlooked, namely the darkness of the eyepiece field.

In 1964 I had the opportunity of using a pair of 20 × 5″ spotter binoculars. These huge binoculars gave a field of 3° and, with a magnification of 20×, should have been far more efficient for comet observing than my own 20 × 65 mm. Indeed, star clusters and the like showed up beautifully in the large instrument; the beautiful pink shade of the central condensation of 47 Tucanae being spectacular. However, a dim comet was visible low in the northwestern sky (Everhart's, about ninth magnitude and very diffuse at the time) and, to my surprise, this was more readily observed in my binoculars than in the large ones—even though a faint star near the coma, barely visible in the smaller instrument, was clear in the larger. In other words, magnitude estimates of the comet would have been higher (by about one magnitude) in the smaller binoculars. Admittedly, my instrument was then quite new, but the full cause of the discrepancy seemed to be found elsewhere. The large instrument gave a very bright field in comparison with the smaller, and the comet's image simply melted away into the background illumination through lack of contrast.

In choosing a pair of binoculars then, it must be remembered that a smaller pair giving a dark field is to be preferred to a large pair if the field of the latter is bright. A useful guide in determining the degree of sky brightness is supplied by the formula

$$x = \left(\frac{A}{M}\right)^2$$

where A is aperture size in millimeters and M is the magnification.

If x has a value of around 50 or higher, the binoculars will have a field sufficiently bright to enable their use as "night glasses" (as, for instance, 7×50 mm which are often employed as "night glasses"); however, for comet-sweeping, the lower the value of x the better.

Thus the $20 \times 5''$ binoculars (20×125 mm) have $x \cong 39$ whereas the 20×65 mm have $x \cong 10.6$.

In sweeping for comets, a simple altazimuth mount is to be preferred; the instrument must be moved horizontally very slowly, and a careful watch must be maintained all the time lest any dim nebulous object should creep into the field of view. This is followed by an overlapping sweep in the opposite direction, the telescope having been raised or lowered slightly.

Comets may appear anywhere in the sky, although statistics show that most reach discoverable brightness, initially, when within 90 degrees west of the Sun.[1] Therefore, the eastern sky before dawn is the region most likely to yield new objects and is thus the prime hunting ground for would-be comet discoverers, although the west after full Moon should certainly not be neglected (indeed, the western evening sky at *any* time should not be neglected) as comets may have approached the Sun during the preceding week or ten days but remained unobservable in the bright sky. Sometimes, a comet well above the minimum observable magnitude will be found under such conditions.

Should the comet hunter happen upon some diffuse patch of

light, he must not jump to any premature conclusions. Reference to a good star atlas (preferably the *Smithsonian Astrophysical Observatory Star Atlas*) will probably reveal that he has "discovered" a nebula known for over a century. Then there are the optical "ghost" reflections of nearby bright stars or even streetlamps (these have even fooled experienced comet observers) and, most troublesome of all (in my opinion at least) unresolved groups of two or three very faint stars, which frequently give the appearance of tiny nebulae in low-power eyepieces.

Once these have been eliminated, it is advisable to check the position of the suspected object with that of any known comet. This may sound obvious, and it is indeed presumed that the comet hunter will subscribe to some form of circulars which keep him abreast of new discoveries, but I am here referring to very faint periodic comets, some of which are always present, but which are generally overlooked by amateur astronomers, being, in the main, far too faint for visual observations. Nevertheless, very occasionally, one of these will flare up greatly in brightness (as we saw in Chapter 1) and may be mistakenly reported as a new object, as happened during the great outbursts of periodic Comet Tuttle-Giacobini-Kresak in 1973.

If the suspected object's position does not correspond with that of any known comet and all other possibilities have been eliminated (a careful watch for any movement, relative to the stars in the telescopic field, over the course of an hour or so should clinch the cometary nature of the suspected object) the comet hunter should send a telegram to the Smithsonian Astrophysical Observatory giving the position and magnitude together with a brief description noting such features as central condensation, tail, and daily motion of the comet.

Serious comet hunters will find some form of notification of discoveries an indispensable asset. The British Astronomical Association provides circulars announcing discoveries of comets, novae, etc., in addition to its excellent *Handbook* which, amid a wealth of other astronomical information, provides predictions for all known periodic comets visible during the forthcoming year. Another service is the International Astronomical Union Telegram Bureau. This circulars-and-telegram service is quicker and more comprehensive, but it is also more expensive.

The discovery of a new comet is an exciting event for those of us attracted to the subject. We immediately inquire: will it become bright and spectacular; is it really a new comet or the return of an object seen long ago; will it pass very close to the Sun or Earth?

Such questions cannot be answered until we determine the comet's position in space and how it is moving relative to the Sun and Earth, and to determine this, an orbit needs to be calculated.

Like all objects in the solar system, comets describe orbits around the center of gravity of the system, which may for most practical purposes be considered to coincide with the Sun, although in actual fact it sometimes lies outside the Sun due to the gravitational effects of the planets (principally Jupiter). The possible paths a comet, or any other body moving in space, may take are straight lines, hyperbolas, parabolas, ellipses, and circles, arranged in order of decreasing eccentricity. In actual fact, we can quickly eliminate straight lines, parabolas, and circles, as these are very special cases unlikely to exist in nature for any length of time. We are thus left with hyperbolas (open curves) and ellipses (closed curves) as the only two viable possibilities. These various curves are illustrated in Figure 7. For a comet moving in a closed curve (ellipse), the gravitational attraction of the Sun will ensure that the comet returns again and again, whereas those objects moving with hyperbolic velocity have attained sufficient velocity to escape the Sun's field and (unless they again become elliptical at great distances, under the gravitational influence of the solar system as a whole) will be lost to the system.

A diagram of a comet's orbit, showing the "elements," is given in Figure 8. In this example, an elliptical orbit is indicated with $e < 1$. For a parabola it is $e = 1$, and for a hyperbola $e > 1$. The corresponding values of a are finite for an ellipse and infinite for the parabolic and hyperbolic cases.

The raw material with which a computer of such an orbit must work is, of course, that provided by positional measurements of the comet on the celestial sphere. At least three sufficiently accurate positions are required (and are sometimes, by the way, supplied by amateur astronomers) spread over a tolerably long arc. Normally, those comets which are sufficiently bright to be discovered visually are fairly close to the Sun and Earth and, conse-

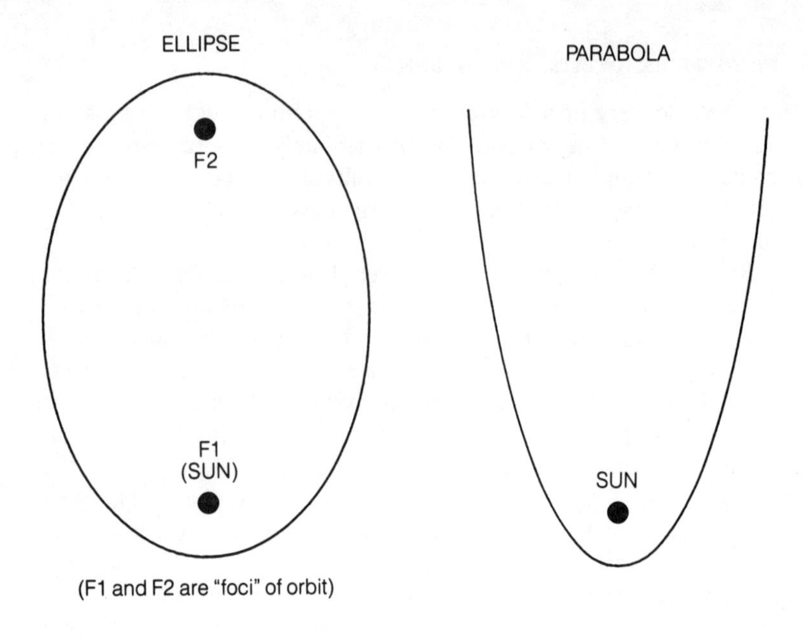

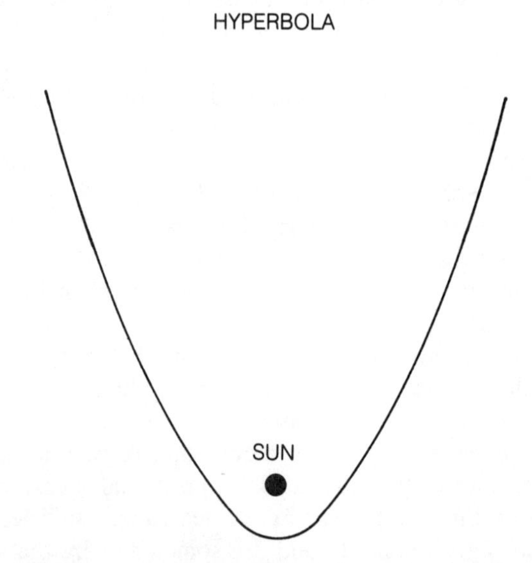

Fig. 7. *An ellipse, parabola, and hyperbola. Only the ellipse and hyperbola are actually realized in nature, although most comet orbits differ only slightly from a parabola.*

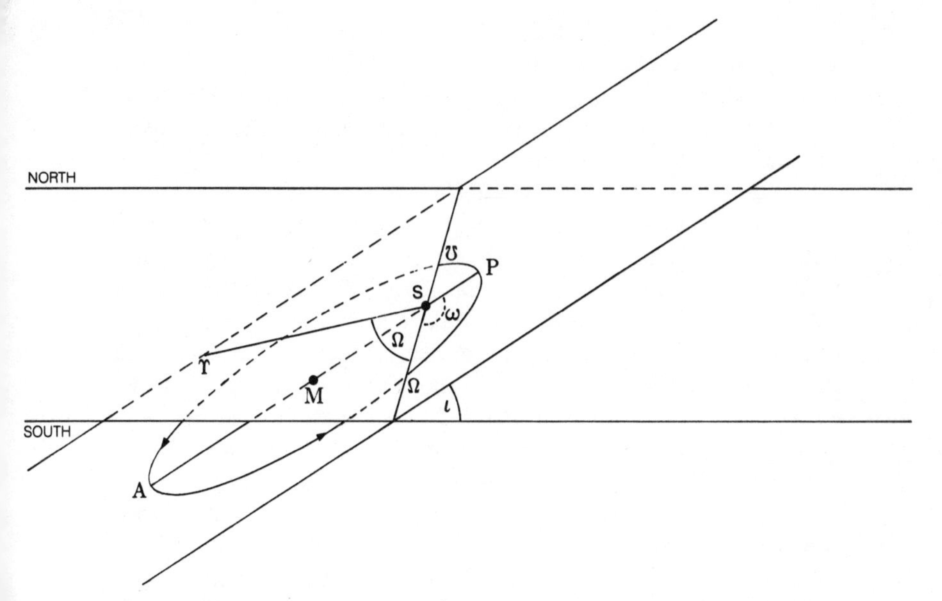

Fig. 8. *The orbit of a comet showing some of the "elements." The units of the distances referred to are given in astronomical units. (1 A.U. = 149,500,000 km, the mean distance of Earth and Sun.)*

S = *Sun* = F_1.
♈ = *First point of Aries.*
M = *Center of orbit.*
P = *Perihelion.*
A = *Aphelion.*
☊ = *Ascending node (point of intersection between the plane of orbit and the plane of the ecliptic—comet moving north).*
☋ = *Descending node (point of intersection between plane of orbit and plane of ecliptic—comet moving south).*
Ω = *Longitude of ascending node (angle from* ♈ *to ascending node, measured in plane of ecliptic).*
ω = *Argument of perihelion (angle from ascending node to perihelion, measured in plane of orbit in the direction of motion of comet).*
i = *Inclination.*
a = PM = AM = *Semimajor axis*
e = $\frac{SM}{PM}$ = *Eccentricity*
q = PS = a (1 − e) = *Perihelion distance*
q′ = AS = a (1 + e) = *Aphelion distance*

quently, their motion across our skies will be rapid enough to describe such an arc in two or three days. Thus, three positions on as many consecutive nights will be sufficient to enable a tolerably accurate orbit to be derived.

For the sake of simplicity, this initial orbit is assumed to be parabolic—not that anyone seriously believes it to be parabolic in actual fact, but it is likely to be an ellipse of such high eccentricity (or, less likely, a hyperbola of such low eccentricity) that parabolic motion may safely be taken as a fair approximation. Eventually, as more positional measurements extending over a wider arc are taken into consideration, the comet's motion may be seen to diverge from that derived on the basis of the parabolic assumption; that is to say, positional measurements will diverge farther and farther from those predicted on the assumption of a parabolic orbit. Thus a new orbit—elliptical or hyperbolic—will need to be calculated.

Some comets are so poorly observed, or have orbital eccentricities so close to the parabolic limit, that their orbits over the observed arc remain indistinguishable from parabolas within the limits of observational error. A smaller number are ellipses of relatively low eccentricity and a still smaller number are definitely, though slightly, hyperbolic, although many of these latter can be shown to become elliptical at great distances from the Sun and all have been demonstrated to have been (within the limits of observational error) originally elliptical. No comet has ever been found to have entered the solar system on a hyperbolic path—an observation which has important implications for cometary cosmology, as we shall see later.

Such "definitive orbits" are not normally calculated until after the comet has faded from view and all the positional information has been published. Not until then is it usually possible to ascertain which comets are moving in elliptical orbits (usually with "periods" of thousands of years!) and which ones may be making their only trip to the inner solar system. When gravitational perturbations by the planets and nongravitational effects are taken into consideration, the orbit can be calculated to a high degree of accuracy, and if the period is sufficiently short to make it a useful exercise, a prediction for the subsequent return is possible.

Early orbit computers, shortly after the comet has been discovered, cannot afford such luxuries, however. Fortunately though, they do not require them, as their work will be satisfactorily accomplished if a preliminary orbit of sufficient accuracy to enable calculation of a useful ephemeris—a type of timetable giving the comet's position on the celestial sphere in right ascension and declination—for the next few weeks, can be calculated. Such an ephemeris will usually contain information on the comet's geocentric and heliocentric distance, apparent magnitude (frequently including the formula by means of which this was calculated), and in those cases where the comet passes at small angular separation from the Sun, its elongation. A typical preliminary ephemeris is reproduced in Figure 9.

As we see it from the Earth, a comet's motion across the sky depends not only upon its actual motion through space, but also on its perspective as seen projected onto the sky. For instance, if the comet is moving more or less toward or away from the Earth, its daily motion will be small and the object may more or less "hover" in the one region for days or weeks. On the other hand, if we are viewing the comet more or less perpendicular to its direction of motion, its movement across the sky may be very rapid —daily motions of six degrees or more are not uncommon. When Comet Suzuki-Saigusa-Mori (1975 X) passed the Earth at the unusually close distance of about 0.1 A.U. in October 1975, its motion was so rapid that it could be detected in only one or two minutes and was relatively conspicuous in only ten or fifteen minutes, although this is, of course, not really typical.

Nevertheless, it is not merely perspective which gives comets varying speeds through the sky. Their intrinsic velocities differ as well, according to Kepler's second law (see Figure 10). According to this law, the time taken for a comet (or other object) to sweep out the arc *CD* is equal to the time taken to describe the much larger arc *AB* in the vicinity of the Sun. For an object moving in an ellipse of very low eccentricity (such as the Earth or any other normal planet), the difference is not very great, but for a comet the difference can be very considerable. As will be readily appreciated, orbits having very small perihelion distances will be especially affected, and it is indeed possible for a comet to be moving at over one million kilometers an hour at perihelion and at little more than walking speed at aphelion!

T = 1979 July 23.405 E.T.
q = 0.3925
ω = 51°.91
Ω = 164.75   epoch 1950.0
i = 134.89

1979 ET	R.A.		Dec.		Δ	r	m1
June 26	8 hours	37.9 minutes	+0 degrees	32.8 minutes	1.180	0.794	10.4
July 1	8	39.32	+6	13.1			
6	8	39.26	+11	35.4	1.293	0.602	9.4
11	8	37.23	+16	47.5			

$m1 = 11.0 + 5 \log \Delta + 10 \log r$

T = time of perihelion passage, in decimals of a day (ephemeris time)

Δ = geocentric distance of comet

r = radius vector of comet (heliocentric distance of comet)

(Distance in Astronomical Units)

Fig. 9. *Sample of an initial cometary ephemeris, that of Comet 1979*c *Bradfield, published on July 2, 1979 (only eight days after the comet's discovery) and is based on four accurate positions obtained June 25–29.* (I.A.U. Circular No. 3375, elements calculated by M. P. Candy from positions obtained by D. Herald.)

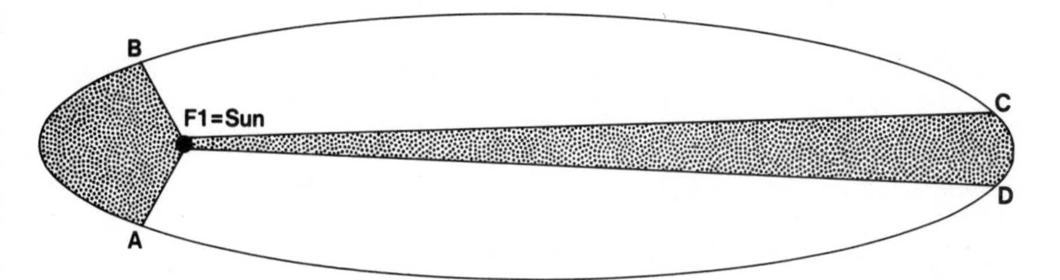

Fig. 10. *According to Kepler's second law, the area* AF_1B *equals the area* CF_1D. *The time taken for a comet to describe arc* AB *will equal the time taken to describe arc* CD. *Therefore the velocity will be much greater at* AB *than at* CD.

Of course, only the near-perihelion portion of the orbits of such comets are sufficiently close to Earth to allow observation; the few comets which can be seen right around their orbits are necessarily short-period objects moving in low-eccentricity ellipses. Nevertheless, the changes in velocity of an object of small perihelion distance does become apparent, with objects observed near the Sun invariably having much larger daily motions than objects with equal geocentric, but much greater heliocentric, distances.

Unlike the planets, comets of very long period show little regard for the plane of the ecliptic. These far-flung members of the Sun's family may approach from any direction, in orbits having any inclination. About half are retrograde—i.e., they move in the opposite direction to the planets, with orbits inclined at greater than 90 degrees to the plane of the ecliptic. The inclinations of some typical comet orbits are represented in Figure 11.

Perihelion distances—or should I say *observed* perihelion distances, as we can only deal with these—range from 0.0048 A.U. (Thome's Comet, 1887 I) to 6.8796 A.U. (Schuster's Comet, 1975 II) and 6.0196 (Van den Bergh's Comet, 1974 XII). These latter objects have by far the largest perihelion distances observed to date and must, surely, represent a multitude of similar objects which forever remain too faint to be discovered. At discovery, these two comets were of magnitudes 15 and 17 respectively, giv-

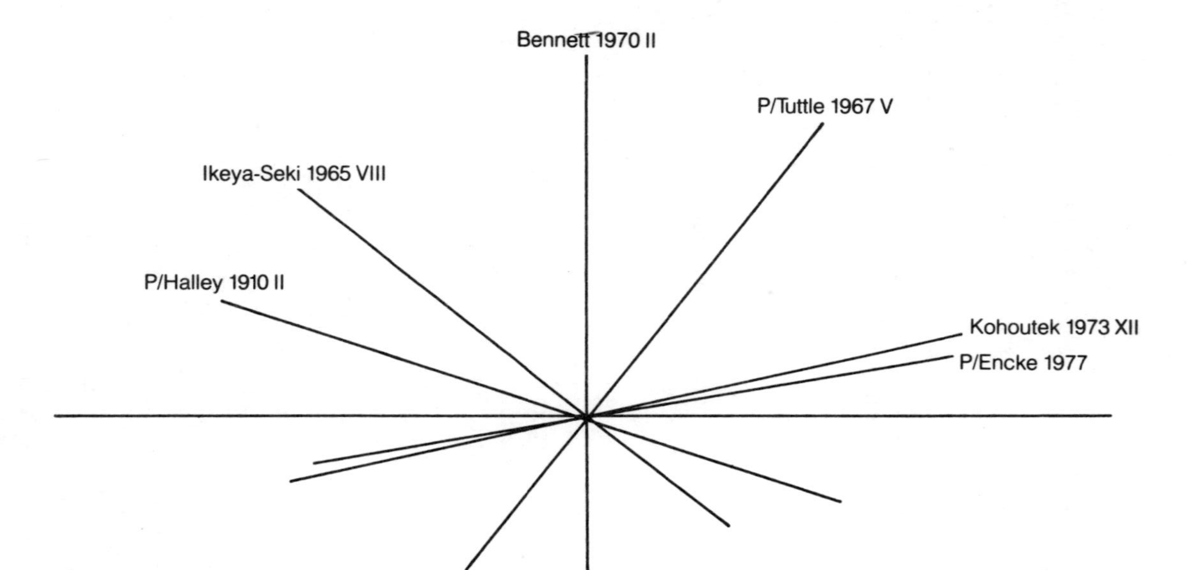

Fig. 11. *The inclinations to the plane of the ecliptic of the orbits of some typical comets.*

ing reduced brightnesses of about 4.5 and 6 or 7 at $r = 1\ AU = \Delta$, assuming that their absolute magnitudes would have remained the same at smaller heliocentric distances. The absolute magnitude of Van den Bergh's Comet is not particularly high, but it is possible that this comet became active at distances where many others of similar "potential brightness," as it were, remain inert and unobservable. In any case, there can be no real doubt that many comets pass at perihelion distances equal to and greater than these two objects, but remain forever invisible from Earth. In fact, there is probably a spherical halo of comets surrounding the Sun, containing vast numbers of objects and being rather analogous to the halo of globular clusters and population II stars surrounding the galaxy. Carrying this analogy further, the solar planetary system is a disk like the population I disk of the galaxy and is also penetrated, sometimes, almost to the system's center of gravity by Sun-grazing comets in a manner strikingly reminiscent of the penetration of some population II stars to near the galactic center. Furthermore, because of the distribution of orbits of both population II stars and (as far as we can tell) of comets, the net angular momentum of both systems is zero, in contrast to the disk components where population I stars, in the one instance, and planets in the other, share the same direction of motion, giving the system a positive net angular momentum. Dynamically, the solar system is a model of the galaxy!

PERIODIC COMETS

When we turn from the near-parabolic objects to those with periods of 200 years or less, we find the stamp of order governing the planetary system increasingly applied, the orbits of most such comets being inclined less than 45 degrees to the plane of the major planets. Of all the known comets of short or moderate period (i.e., less than about 200 years) the only ones having retrograde motion are Halley's, Tempel-Tuttle, Pons-Gambart, and Swift-Tuttle. It is noteworthy that none of these comets has a period of less than 30 years; all those of *very* short period have much lower inclinations.

About 110 such comets are known—including those which

have been seen at a single return only. They are (with the few exceptions discussed earlier) known by the name of their discoverer(s) together with the prefix "P" to designate periodicity. If one person has more than one periodic comet to his credit, a number is also added as in P/Brooks 1 and P/Brooks 2, or P/Schwassmann-Wachmann 1, P/Schwassmann-Wachmann 2, and P/Schwassmann-Wachmann 3.

Short-period comets move in orbits ranging from near circles to fairly elongated ellipses; however, many have their aphelion points at heliocentric distances approximately equal to those of the great planets. Thus, those with periods of from 5 to 12 years have aphelia near Jupiter's distance, those of 13 to 18 near Saturn's distance, those of about 28 years near Uranus', and those of from 49 to 81, near Neptune's. For those comets with periods of 100–150 years, a hypothetical planet beyond the orbit of Pluto was postulated.

The cometary members of these "families" were believed to be either long-period objects captured by the great planets or (a less popular theory) ejected by the planets themselves.

Further investigation failed to support this alleged connection with the great planets except in the case of Jupiter's "family" and, perhaps, one or two other comets. In general, although the comet's aphelion may be at the same distance as the orbit of the planet in question, the orbital inclination of the comet is such that it never comes near the planet itself. For instance, P/Halley, although allegedly one of Neptune's family, has an orbital inclination so great that it cannot approach Neptune any closer than 8 A.U.! This comet, and others like it, approach much more closely to Jupiter, however. Indeed, virtually all periodic comets seem to be, to a greater or lesser extent, under the influence of that planetary colossus.

Sometimes the influence of Jupiter is very marked, as is evidenced by the case of P/Wolf, now a classic in the history of planetary perturbations. This comet was discovered in 1884 as a fairly easy telescopic object. However, calculations by Kamienski revealed that before 1875 it had moved in a much larger orbit and would have been quite beyond the instruments of the day. In that year, though, it passed within 0.12 A.U. of Jupiter and had its perihelion distance decreased from 2.54 to 1.59 A.U., thus making it a much brighter object and bringing about its discovery. The comet again passed near Jupiter in 1922, whereupon its orbit reverted back almost to the pre-1875 position and, consequently, it is now once more a very faint object visible only in very large telescopes.

Many other periodic comets (e.g., Brooks 2, West-Kohoutek-Ikemura, or Faye—to name but a very few) have been discovered after close approaches to Jupiter, whereas others (e.g., Lexell and Oterma) have been lost after having their orbits altered by this planet to such an extent that they no longer come sufficiently near the Earth and Sun to be observable. (Lexell's orbit, for instance, which was less than six years, was extended to about 260 years, with the aphelion of the old orbit becoming the perihelion of the new. There is even a chance that the new orbit may be even more eccentric than this, and a hyperbola is not beyond the bounds of possibility.)[2]

In addition to such large and dramatic changes, perturbations by Jupiter can cause persistent small orbital alterations which, over a period of years, can add up to considerable differences in a comet's orbit. A striking example of this effect is provided by P/Pons-Winnecke, first observed in 1819 and rediscovered by Winnecke in 1858. Since then, it has been observed at most returns and has a period of about six years. Yet, the orbit followed today by this comet bears little resemblance to that of a century ago. So great have been the effects of Jovian perturbations that the inclination of the orbit has more than doubled and the perihelion distance increased from 0.77 A.U. in 1819 to 1.25 A.U. in 1976. Such alterations as these could easily result in a comet becoming lost, especially in the days before high-speed electronic computers.

Thus Jupiter (and, to a lesser extent, the other giant planets) may be seen as a great disturber of the cometary peace within the solar system. Yet, in playing this role, the stamp of order is applied to the central cometary system—the inclinations of short-period orbits are kept low (by cometary standards) and any comet which may be thrown into a retrograde orbit runs the risk of meeting Jupiter head on at a subsequent return and of consequent acceleration into a hyperbolic orbit.

Long-period comets having their perihelia near the orbit of Jupiter, inclinations close to the plane of the planet's orbit, and motion nearly parallel to that of Jupiter's are the ones most likely to have their orbits reduced (through several revolutions) into the short-period variety (although a *very* small percentage of such comets are actually captured). Those meeting Jupiter head on or those having small perihelia, passing the planet on their way to the Sun, are thrown into hyperbolic orbits due to acceleration caused by the planet's gravity. Small-perihelion comets which meet Jupiter *after* passing the Sun are thrown into retrograde ellipses which become hyperbolas when the comet meets Jupiter head on at a subsequent return.

Thus far it has been assumed that the motion of a comet through space can be accounted for by considering only the gravitational attraction of the Sun and planets. For certain comets, however, the situation is not this simple.

In 1818, the mathematician Encke discovered that the fairly faint comets observed in 1786, 1795, 1805, and 1818 were, in fact, returns of the same object having a period of only 3.3 years—a very radical conclusion in those days when only P/Halley was known to be periodic. Nevertheless, the comet proved Encke correct by returning as predicted and has been faithfully coming back ever since.

Having an orbit extending only into the asteroid belt, this comet is free from any serious planetary perturbations, but nevertheless, it quickly became apparent that its times of perihelion passage were always slightly wrong. Irrespective of how accurate the calculations were, the discrepancy still existed. It seemed clear that the comet's orbit was constantly growing smaller—which at least explained one difficulty: how it came to have such a small orbit in the first place!

Early suggested explanations for this phenomenon involved the postulation of some medium through which the comet moved, losing by means of friction some energy every time this medium was encountered, and thereby falling into an ever smaller ellipse.

Such an explanation was not convincing, as such a resisting medium would affect all comets of similar perihelion distance equally—in contradiction of the observed facts. Indeed some comets like P/d'Arrest actually seem to be *increasing* the size of their orbit, and no resisting medium can *add* energy to a comet, as would be required by these observations.

The consensus nowadays is that these "nongravitational effects" arise from the evaporation of cometary ices, the resulting gas emission giving thrust to the nucleus. The ejection of gas will occur on the sunward side and, in the case of a nonrotating nucleus, result in a push in the opposite direction to the Sun.

When the nucleus is rotating (as, presumably, most will be) there will be a delay in the maximum emission in the sense that it will not be at the subsolar point, but rather in the "afternoon" of the nucleus' "day." The thrust will now be at an angle to the radius vector. For those comets whose nuclei rotate in a retrograde direction, the thrust will be such as to decrease their orbits (as is the case with P/Encke). Direct rotation gives the opposite result, as is evidenced by P/d'Arrest (Figure 12).

Nongravitational effects appear to arise from continuous emission of cometary material and are not, apparently, correlated with brightness fluctuations. On the other hand, there does seem to be a correlation between the magnitude of these effects, the quantity of material emitted by the comet, and the mass of the nucleus. This is, of course, only to be expected if the current theory is correct.

Thus P/Giacobini-Zinner shows quite marked nongravitational effects indicative of a small nucleus of rather low mass. Supporting evidence for this supposition is supplied by the feathery, low-density nature of the meteors associated with this comet, suggesting a nucleus composed of very loose material.

Similarly, calculations by B. G. Marsden[3] have shown that the smaller component of P/Biela was more strongly affected by nongravitational influences than the major one. Apparently, this was a very small, icy fragment which became detached from the main mass.

More recently, similar evidence seems to have been supplied by the unstable comet P/Perrine-Mrkos. Always prone to rather large nongravitational effects, this comet was well off course at its 1968 return, when it was also much fainter than had been anticipated. At its next scheduled return in 1975, the comet was not found and may have disintegrated, suggesting that only a very small remnant may have returned in 1968.

This theory of nongravitational effects readily explains the observed changes in magnitude and even in the sign of these effects.

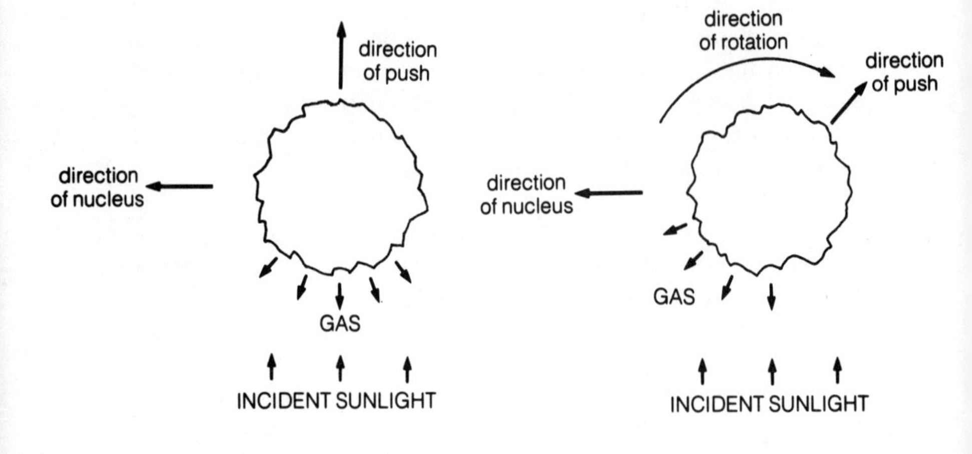

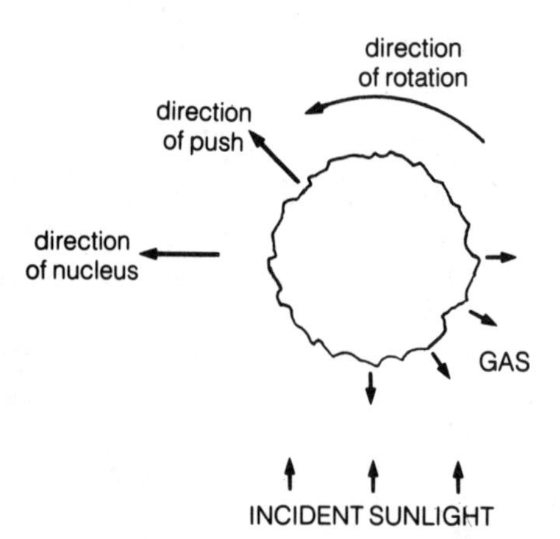

Fig. 12. *Nongravitational effects. The comet nucleus is moving toward the left in each case. Case 1 gives a nonrotating nucleus, case 2 is retrograde rotation, and case 3 is direct rotation.*

It has long been recognized that the changes in the orbit of P/Encke are less pronounced now than they were last century, whereas other comets (e.g., the now-lost comets P/Brorsen and P/Biela) were observed to experience increasing nongravitational effects. Originally, the variation in nongravitational effects experienced by P/Encke was explained in terms of a relatively nonactive core being gradually exposed by the evaporation of a volatile mantle. Such a model is, indeed, supported by the dense nature of the meteoroids associated with this comet (suggesting that they were released from the central regions of what was once a large nucleus). Conversely, the increased acceleration of P/Brorsen and P/Biela would be understandable if these comets were icy throughout, with the effects of thrust increasing as the residual mass decreased (a hypothesis which is supported by the total disappearance of these two comets).[4]

Nevertheless, although the general reasoning of this hypothesis is probably correct, the example of P/Encke now appears somewhat more complex than this. Recent work by Whipple and Sekanina indicates that the decrease in this comet's nongravitational parameters is the result of precession of the spin axis of its nucleus. This precession is interpreted as arising from the jet forces of subliming ices acting upon the oblate-spheroidal nucleus.[5]

The phenomenon of precession is also able to explain the reversal of sign in the nongravitational effects operating upon some comets. If precession of the axis is sufficient, the sense of rotation can go from retrograde to direct or vice versa. Thus P/Faye has changed from acceleration to deceleration whereas P/Borrelly has gone from deceleration to acceleration.

THE ORIGIN OF COMETS

Leaving aside the very early, "meteorological," speculations of Aristotle and his followers, theories about the origin of comets may be divided into two broad groups, namely those which postulate an interstellar origin and those which postulate a planetary one.

Interstellar Origin

The foremost exponents of the interstellar theory of comets were Laplace, Von Seeliger, Fabry, Bobrovnikoff, Nölke, and more recently, Lyttleton.

With the singular exception of Lyttleton, these theorists took as the starting point of their hypotheses, the prior existence of condensations within interstellar clouds. The differences in their respective theories mainly concern the manner in which such condensations became incorporated into the solar system as comets.

Laplace, Von Seeliger, and Fabry believed the Sun to move in a general, permanent, interstellar field of comets, its gravitational force attracting those members of this field whose motion relative to the Sun is nearly zero.

Bobrovnikoff[6] extended this theory somewhat by pointing out that from his own research on the lifetimes of comets, the objects now associated with the solar system must have been captured during the last million years. Within this time, the Sun must have passed through a cloud containing such cometary condensations resulting in massive captures. No mechanism is suggested, by Bobrovnikoff, through which such a capture process might take place.

Nölke (1936)[7] attempted to prove that a "Bobrovnikoff-type" capture could occur if a resisting medium was postulated. Such a medium, Nölke believed, would be provided by the general interstellar field of dust and gas.

A major objection to all such theories of the capture of interstellar comets is the lack of definite hyperbolic orbits among observed new comets. If comets are, indeed, captured from interstellar space, it is remarkable that none has ever been observed *entering* the inner solar system on a hyperbolic path. Those with *slight* hyperbolic velocities near perihelion can all be shown to have acquired such motion through planetary perturbations, as discussed above.

Of course, if the process of capture was completed long ago, this need not be an objection, as the chances of an interstellar stray (assuming such objects to exist) arriving at the neighborhood of the Sun sufficiently closely to be observed by earth-bound astronomers is not very great, and the consequent lack of strongly hyperbolic orbits is no longer a great difficulty.

Moreover, one may expect that the general circumsolar system of comets would show some distinct preference for the galactic plane, if these objects were really captured from interstellar space. No definite preference has been found, however, which cannot be more readily explained away as a selection effect.

Thirdly, such theories as these do not really explain how comets came to be formed in the first place, only how they came to be incorporated into the solar system.

The first and last objections at least are avoided by the theory of Raymond Lyttleton, which forms, as it were, a bridge between truly interstellar and truly planetary theories, and must be regarded (at least from the methodological point of view) as the most philosophically sophisticated cometary theory yet postulated. Lyttleton attempts not merely to explain the origin of comets, but also to account for their observed properties and behavior, in a single, comprehensive, theory. Furthermore, his cometary hypothesis is a corollary of the general theory of accretion of interstellar material by the gravitational lens effect, thereby bringing cometary cosmology into the mainstream of astrophysics and attempting a unified theory relating the seemingly diverse phenomena of stellar formation and cometary activity.

Lyttleton considers a star (e.g., the Sun) entering an interstellar dust cloud, and he demonstrates how the particles of the cloud will describe hyperbolic paths around the Sun, with these paths intersecting along the line passing through the Sun and parallel to the relative velocity vector of the Sun and cloud. As seen from the apex of solar motion, these points of intersection lie behind the Sun, resulting in a large number of collisions of particles and an accretion of material along this line. Inelastic collisions will give rise to a "high-density" (in relative terms, of course) stream along the accretion axis; and this, in turn, will break up into discrete clouds of particles. It is these small, discrete clouds which Lyttleton identifies with comets, and he then proceeds to explain how such a comet model accounts for the observed properties of these objects.

Thus each of the tiny particles is describing its own orbit around the Sun, unaffected by the minimal gravity of neighboring particles or the cloud as a whole. When far from perihelion, the cloud is large and diffuse, but as each particle nears a (more or less) common perihelion, the swarm contracts and collisions between individual particles become more frequent. Fine dust and puffs of gas resulting from such collisions, according to Lyttleton, supply material for the coma and tail.

The process of accretion in general has been the subject of much discussion and seems to have a good deal in its favor. Still, the exact process as discussed by Lyttleton is said to be extremely sensitive to the relative motions of the star and dust cloud, and may also be very sensitive to other factors such as turbulence within the cloud.

Nevertheless, even if the entire process is valid, and even if the Sun is passing, or has recently passed, through such a cloud (which is not impossible, as a diffuse cloud would not have any pronounced effects upon the Earth), and even if small discrete clouds do actually populate the solar system, it does not necessarily follow that Lyttleton's theory of comets is thereby validated. Another early worker in the field, H. Bondi, left open the question as to whether such "clouds" could be identified with the comets.

Certain considerations do, indeed, throw considerable doubt on such identification. For instance, the Lyttleton theory, as it stands, seems unable to adequately explain the following:

1. Large meteoroids in cometary meteor showers. This has already been considered in our discussion of the cometary nucleus. Sufficient here to remember that evidence of meteorites sometimes falling from showers such as the Taurids and fireballs as bright as the full Moon in the Leonids, runs counter to a model in which comets consist *entirely* of very small dust particles, as necessitated by Lyttleton.

2. Lyttleton's model would suggest that comets with a higher than usual dust content should also be strong producers of gas, due to the greater number of collisions which must occur within

such an object. However, spectroscopic analysis of the two bright comets of early 1970 showed similar quantities of gas emission in both, despite the fact that the first (Tago-Sato-Kosaka) was relatively dust-poor, whereas the second (Bennett) was particularly rich in dust. Analysis of these two objects from Earth-orbiting satellites also revealed that comets are even richer in gas than had hitherto been believed, an observation adding yet another heavy burden on any theory which requires the cometary gas content to be supplied by collision of small particles of dust. Even before these observations were made, the observed gas content of comets placed a considerable strain upon the theory.

3. According to the theory of Lyttleton, comets with perihelia as small as those of 1963 V, 1965 VIII, and 1970 VI would be totally vaporized at, and near, perihelion. Even an object composed of particles the size of large gravel could not survive such an encounter.

This much is recognized by Lyttleton, who then goes on to argue that even a gaseous comet would remain relatively stable for the duration of its fiery loop around the Sun and would then recondense as it traveled out again into the cooler regions of space.

The major drawback to this attempted solution is simply that the process would turn the entire comet into a body as sensitive to the Sun's radiation as a normal cometary tail. That gas is blown away from comets near the Sun is undeniable; it is the main characteristic of these objects, and very good reasons would need to be offered in order to explain why the *entire* comet should not be blown away into space in the form of a gas tail, if all its matter was converted into gaseous form. Similarly, even assuming for the sake of argument that enough of the comet remained (after hours near the Sun in gaseous form) to recondense into dust particles (itself an unlikely process under such conditions), it is surely relevant to note that such particles would, for a certain length of time during the early stages of their regrowth, be of the order of size of typical dust-tail particles. Why would *these* particles not be blown away by the powerful radiation of the nearby Sun, when similar particles are repelled by solar radiation at heliocentric distances greater than the planet Jupiter?

Although strongly critical of "evidence" supplied by observations conveniently altered to support one's own theory, Lyttleton (unintentionally, one presumes) commits that very transgression in dealing with the breakup of the Sun-grazing comet 1882 II. He states that shortly after perihelion, the nucleus of this comet "elongated" into a brilliant streak[8] and subsequently divided into several discrete subnuclei. Such a process is, he believes, quite consistent with a totally gaseous comet and is dynamically similar to the hypothetical preplanetary tongue of gas, drawn out of the Sun, as postulated by James Jeans in his theory of the formation of the solar system.

Indeed, expressed in terms of "elongation," the behavior of 1882 II does seem to suggest the development of a fluid, gaseous body, although it must still be borne in mind that some Sungrazers did *not* (as far as our observations show) similarly divide near perihelion and that the process cannot, therefore, be considered inevitable even for an allegedly totally gaseous comet (as, one would think, it would be if Lyttleton is correct).

Examination of contemporary reports of Comet 1882 II, however, does not support Lyttleton's account of the phenomenon, and consequently, his explanation begins to look doubtful to say the least.

Apparently, the nucleus remained a single object until September 28, eleven days after perihelion—by which stage the comet had receded to a distance of 0.57 A.U. from the Sun (a distance at which, according to Lyttleton's theory, its component particles should have well and truly recondensed). By September 30, two elongated nuclei were visible, and up to five were observed in early October, at times connected by a bright wisp of nebulosity, earning the formation the name of "the string of beads." This "string" undoubtedly was what suggested "elongation into a bright streak" to Lyttleton, but as we can see, the comet was at planetary distances from the Sun when this phenomenon appeared and some other explanation must be considered.

A repeat performance of this general phenomenon was staged in 1965 when Comet Ikeya-Seki (1965 VIII) followed a path almost identical to that of 1882 II and displayed very similar phenomena. This comet was kept under observation throughout its passage through perihelion, and although displaying instability and throwing off fragments, it at no time gave the impression of being totally fluid. Quite the contrary, it appeared to be solid throughout. (By way of an aside, it has been remarked that the separation of fragments so close to the Sun—if not actively supporting the Lyttleton model—at least casts doubts upon the rival icy-conglomerate model as fragmentation should expose fresh ice and result in sharp outbursts of brightness. However, at this distance from the Sun, the "normal" increase in brightness is likely to be greater than a typical outburst, and any increased activity which may occur due to schism is likely to be "swamped.")

After perihelion, the nucleus remained visible as a single discrete object for a few days, after which it became unusually large and diffuse. At this stage, the comet had receded to nearly 0.3 A.U. from the Sun. In early November, this enlarged nucleus elongated (though not into a "brilliant streak") and split in two, the minor component itself being a very close triplet of sub-sub-nuclei. Division of a fluid body does not seem able to explain these events. Rather, the view put forward by Sekanina[9] looks more promising—i.e., that the (solid) nucleus suffered a tidal disruption very close to perihelion. Separation of the two fragments would have been quite slow, and it is not surprising that the double nucleus could not be observed initially, as such, but as the two components separated, they first appeared as an unresolved (and therefore elongated) double and finally separated sufficiently to be observed as two objects.

4. Comets forming in the accretion stream will fall toward the center of gravity of the solar system. Depending on the position of Jupiter and the other great planets, the center of gravity of the solar system can vary (relative to the center of the Sun) from zero separation to about 2.20 solar radii.[10] Thus a comet falling toward the center of the system would either hit the Sun or pass into a Sun-grazing encounter, during which time it would be (according to Lyttleton) totally vaporized.

Aside from the difficulties already discussed pertaining to Sun-grazing encounters, such a theory is hard-pressed to explain the number of comets having very large perihelia. The early 1970s witnessed a rich harvest of such objects, and as there can be little doubt that these observed comets are only the brighter representatives of their class, we are practically forced to admit that most comets have perihelia lying well outside the orbit of the inner planets. If this is doubted, just examine the list of comets discovered between 1970 and 1975, and note the frequency of those coming within the Earth's orbit and those lying well without. Admittedly, distant objects are slower moving and thus have more time in which to be found, but they are also very faint and surely must escape detection much more readily than those of small perihelion distance.

Conversely, it seems that Sun-grazing comets are exceedingly rare. As will be seen later, there is evidence suggesting that all known Sun-grazers may be fragments of a single, disintegrated comet, and therefore even their comparative frequency during the last few centuries may be somewhat deceptive—we may have only been seeing the fragments of *one* freak comet masquerading as individual comets in their own right.

It would seem, by virtue of the above considerations, that any observation explicable using Lyttleton's theory is more readily explained by the icy-comet model. It remains true, however, that Lyttleton's theory is, philosophically, a very sophisticated one in that it attempts to explain *all* cometary phenomena from a single set of initial, strongly supported presuppositions. This has such a great aesthetic appeal that it is hard for some people to accept its total inaccuracy, even though any theory (even the most aesthetically pleasing) must stand or fall by empirical evidence.

Nevertheless, it does seem that the weight of empirical evidence is against the comet model which follows from the theory. Is it possible that something has been overlooked, and that some form of accretion process might yet be evoked to explain the formation of a more realistic cometary structure?

Planetary Origin

a. *The Theory of Vsekhsvyatsky*

Vsekhsvyatsky continues a line of thought, expressed by Lagrange (1814), Proctor (about 1870), and Tisserand (1890), which sees comets as objects ejected from major planets by some extremely violent form of volcanism.[11] Observational evidence for such a conjecture is not supplied in any direct manner; no comet or cometlike body has ever been seen emerging from a planet.

Vsekhsvyatsky points to the short lifetimes of periodic comets as the main support for his theory. Such comets, according to his research, typically fade noticeably over a period as short as three or four revolutions of the Sun. After a few centuries, or even a few decades, they must, he argues, disappear completely. Indeed, so rapid is their disintegration that had there been large telescopes in Halley's day, most of the faint short-period comets observed with such hypothetical telescopes would, by now, have faded and disappeared.

The fact that we see periodic comets today must surely mean that new objects are forever appearing in the place of those which melt away. But, if the lives of short-period comets are as short as Vsekhsvyatsky maintains, the chance capture of long-period comets by Jupiter is not sufficiently frequent to restore the balance and account for the number of periodic comets still visible. From whence, we may indeed ask, do these short-period comets come?

The answer which seems most likely to Vsekhsvyatsky, is that they come from the great planets or, in a later development of his theory, from the satellites of these planets. He originally maintained that even the terrestrial (i.e., Earthlike) planets may well have experienced a phase of volcanism sufficiently violent to eject comets and minor planets. Nevertheless, Vulcan's fires have long since died to a flicker of their former glory on Earth, and (if Vsekhsvyatsky is correct) the only places where such violent volcanism is now to be found are on the great planets, Jupiter and Saturn, and (though to a milder extent) on their larger moons.

Vsekhsvyatsky finds evidence for this conjecture in the spots and turbulences observed on these planets, which he interprets as the results of volcanic upheavals of enormous dimension. One would think that Jupiter would be the most active planet, but Vsekhsvyatsky appears to favor Saturn as the solar system's "Valley of Ten Thousand Smokes," on the assumption that the great ring system originated through volcanic activity of a singularly violent variety.

Today, there is little support for the volcanic theory of the rings of Saturn (despite the fact that Vsekhsvyatsky predicted, as a consequence of his theory, a similar ring for Jupiter—which has been discovered by the Voyager 1 spacecraft), and the spots and swirls on the great planets seem meteorological rather than volcanic. Thus his emphasis has shifted rather more to the satellites of these planets as the birthplaces of comets.

Of course, these smaller worlds can hardly be expected to sustain activity of the magnitude postulated for the great planets themselves, but fortunately they are not required to. An eruption the size of Earth's Krakatoa or Tambora would be quite sufficient to eject a comet from these objects of low gravity.

Until recently, few astronomers would take the idea of extensive volcanism on the moons of Jupiter or Saturn seriously. But this has all been changed by the spectacular pictures of Io sent back to Earth from Voyager 1. These reveal a strange world of belching volcanos, smoking fumaroles, and clear evidence of sustained activity beyond anything experienced on our own planet in recent geological times. Moreover, the volcanos of Io appear more violent than those of Earth, and although it is still a vast leap from evidence of volcanism to the observation of an icy comet emerging from this world, I daresay that this recent discovery will give the ejection thesis a new lease on life in the minds of some people.

Be this as it may, Vsekhsvyatsky's evidence for the rapid fading of periodic comets has given rise to much controversy in recent years. Without doubt, *some* periodic comets fade in the course of several trips around the Sun, and there are clear cases of vanishing comets such as P/Biela, P/Westphal, and the strange case of P/Helfenzrieder of 1766. The latter was a naked-eye object with a head as bright as a third magnitude star and a tail in excess of seven degrees in length. Yet, according to orbital calculations, it moved in an ellipse with a period of little more than four years, but it was not seen previously and has not been seen since. Unless the positions upon which this orbit was based were very inaccurate, there must have been something distinctly odd about this comet!

Obvious fading has occurred in the case of P/Faye, which was a dim naked-eye object in 1843 but has since grown steadily fainter until, even at the favorable return of 1969, it barely attained eleventh magnitude.

On the other hand, David Meisel's[12] analysis of brightness estimates of P/Tempel 2 during its exceptionally favorable apparition of 1967 revealed this object to have an absolute magnitude of 9.4, agreeing well with the average of 9.6 for the returns observed between 1873 and 1920, in contradiction to Vsekhsvyatsky's conclusion. More recently, analysis of P/d'Arrest's magnitude during its 1976 return by John Bortle actually revealed a *brighter* absolute magnitude than a century ago,[13] and a number of other periodic comets returning in recent years have similarly failed to show any evidence of secular fading.

Clearly, if rapid fading is not typical of short-period comets (which is not the same as saying that *no* short-period comet fades rapidly), the need for a mechanism to quickly replenish the stocks of these objects is not required—the slow depletion which does occur can be adequately accounted for by the capture of long-period comets by Jupiter. This is the "orthodox" position and the one most favored by contemporary evidence and theoretical investigations.

One might be excused for thinking that any controversy regarding this matter of cometary fading should be easily settled by merely comparing the absolute magnitudes of a sample of comets at different returns; however, the matter is not, in practice, as easy as this. The main problem is the annoying fact that magnitude estimates arrived at by the use of different instruments can differ quite considerably—especially when the comet is very diffuse (such as a periodic comet close to the Earth, for instance). Last century, estimates were mostly made with the aid of small wide-field telescopes which consistently give brighter magnitude values than the photographic methods employed today. This could hardly account for large diminutions of brightness, as in P/Faye, but it may well account for the smaller changes alleged for some other comets.

Also, as we saw earlier, certain periodic comets are prone to drastic changes in their orbits through the gravitational action of the planet Jupiter, and this may affect their intrinsic brightness (it will, of course, affect their apparent or observed brightness). For instance, P/Honda-Mrkos-Pajdusakova, with a perihelion of less than 0.6 A.U., regularly becomes one of the brighter members of Jupiter's family, often reaching about magnitude 8 or 9 for a few days. Yet its brightness is so strongly dependent upon its heliocentric distance that it has not been observed at distances greater than about 1.5 A.U. Now if this comet were to undergo powerful planetary perturbations giving it a new perihelion distance of 1.5 A.U. or greater, it would appear to us (if we could see it at all) as an intrinsically much fainter object. Perhaps part, at least, of the decrease in the intrinsic magnitude of P/Pons-Winnecke is explicable in this way (see the earlier discussion of the orbital changes of this comet). Indeed, Dr. Fred Whipple once calculated that on the basis of the supposed fading of this object, it should have disintegrated completely by the end of the 1950s. However, the comet's demise has not yet occurred, and P/Pons-Winnecke appeared on schedule in 1976.

Similarly, subsequent returns of periodic comets often reveal these objects under very diverse conditions; on one return a comet may come near the Earth and be seen as a relatively bright object, whereas the next return may place it, at its brightest, behind the Sun, thereby making observation impossible except for a short period months before (or after) perihelion. In the latter instance, the comet is seen when far away from the Sun and is likely to be judged as an intrinsically fainter object than the previous return would suggest.

Thus, in the light of this and other difficulties faced by the ejection theory, we are unable, it would seem, to conclude in its favor —even in spite of the volcanos of Io.

b. *The Theory of Opik, Oort, and Van Woerkom*

J. H. Oort's initial listing of the reciprocals of the semimajor axes of twenty-one long-period comets revealed a very interesting phenomenon. There appeared to be a clear concentration of objects having a semimajor axis exceeding 20,000 A.U. Later research, based upon a larger sample of orbits, revealed the same phenomenon and further investigation gave a range for these comets of between 40,000 and 150,000 A.U., with an average of around 50,000 A.U.[14]

Now, a comet entering the central planetary region of the solar system along such an extremely elongated ellipse will suffer gravitational perturbations by the planets, the most minor of which would be sufficient to drastically alter the semimajor axis of its orbit. It seems, therefore, that the semimajor axes of long-period comets should reveal no such concentrations if these objects have already been subjected to planetary perturbations on a previous return. The obvious explanation of why such a concentration does exist is, therefore, that most of these comets having semimajor axes lying within the specified range are "new" objects coming into the central solar system for the first time. If this conclusion is correct, it seems reasonable to believe that some "reservoir" of comets exists at distances between, about, 40,000 A.U. and 100,000 A.U. and that it is from here that comets come.

(Such objects increase in brightness unusually slowly as they approach the Sun, being brighter at larger heliocentric distances than comets of other classes and thereafter increasing less dramatically in brightness. Nevertheless, Whipple has shown that the rate of fading is statistically the same for "new" and "old" comets. Apparently, the early brightness of "new" objects is associated with richness in volatile ices.*)

Of course, simply showing where comets come from does not explain how they came to be there in the first place. Here, we seem faced with two possibilities, namely, either they were formed out there originally, or they were formed within the central planetary system and subsequently perturbed into the circumsolar shell.

Originally, Oort and Van Woerkom postulated a common origin for comets and minor planets and looked for this in the disintegration of a planet between the orbits of Mars and Jupiter. Opik likewise suggested that comets were formed in the vicinity of Jupiter and were subsequently ejected from the planetary system, although he also, at one stage, suggested an alternate possibility

* Previously, it was believed that "new" comets distinguished themselves spectroscopically by exhibiting a larger dust-to-gas ratio than is usual for comets of other classes. This, however, has now been disproved by B. Donn on statistical grounds. It appears that the spectral dust-to-gas ratio is a genetic property of a given comet, irrespective of whether the object is a "new" or an "old" one, and this feature cannot, therefore, be used as a mark of distinction between these two classes of comets.[15]

—that they were originally formed in the outer regions and possibly represent the remnants of the protoplanetary nebula.

Oort, and also Whipple, have opted for this thesis in recent years. The main problem is to get sufficiently dense condensations in the solar nebula so far from the central region. Merely assuming initial inhomogeneities is one possibility—it may even be the correct one—but it has an unfortunately *ad hoc* appearance.

One possibility is that the comets represent the original planetesimals (the bodies that formed the planets of the solar system) —at least those which formed beyond the region within which actual planetary accretion occurred. Another is the very strong probability that the Sun passed through a T Tauri stage in its evolution. T Tauri stars are believed to be very young objects which have not yet settled down to the relatively orderly existence of their mature brethren. These stars appear to possess very powerful stellar winds, and if the Sun did indeed pass through a period as a T Tauri star, solar winds may have been a million times as intense as those experienced today. Such "solar hurricanes" probably ejected considerable quantities of material into regions beyond the planets, "dumping" it there as the outward-bound hurricane turned into a gale and then into a mere breeze.

Despite some "loose ends," it is undoubtedly true to say that some form of the "cometary shell" theory is the most popular among astronomers today. It seems to explain the facts and is easily married to the Whipple icy-conglomerate theory. Whether the comets *formed* outside the planetary system is a more controversial point, as is the possible common origin of comets and minor planets. As yet, no object like a minor planet has been discovered moving in a nearly parabolic orbit, and any speculation concerning a possible circumsolar halo of minor planets beyond the orbits of the outer planets is pure conjecture at this stage.

c. *The Theory of Ovenden and Thomas Van Flandern*

Recently, a new version of the theory that comets and minor planets have a common origin has been put forward by Ovenden and developed by Van Flandern. The distinctive features of this theory are its rejection of a circumsolar cloud as such, the postulation of a Jovian planet as the parent body, and (most controversial) a geologically recent date for the disintegration of this planet—a mere five million years ago.

Van Flandern[16] bases his theory on orbital distribution of a sample of comets having very long periods (corresponding to Oort's "new" objects). He argues that if these orbits are the results of random perturbations from a circumsolar cloud (passing stars, according to Oort's theory, act as the main deflectors of comets into the inner solar system), they should show a more nearly random distribution than they in fact do. Van Flandern argues that the distribution he has found can best be explained if these comets represent "debris" from a great explosion between the orbits of Mars and Jupiter. Furthermore, the fact that the orbits of these objects are not random indicates that the comets are making their first return to perihelion after the event—perturbations by the planets, especially Jupiter, would have altered many of these orbits after only one revolution and destroyed the strongly nonrandom pattern. Thus their periods of revolution should be equal to the time which has elapsed since the catastrophe, and as all the cometary periods in the sample are about five million years, this (according to Van Flandern) is the age of the cometary system (and the system of minor planets and meteorites as well, as these were all supposed to have formed together, according to this hypothesis).

This hypothesis has been described as "wildcat" by some astronomers,[17] and it is, indeed, radical. All I wish to say here is that it seems very hard pressed to explain asteroidal scars of great antiquity on the Moon, Earth, and other inner planets. Most of these are of great age, and if comets and minor planets were not around when they were caused, then what class of object *did* cause them?

Similarly, where is the evidence that the inner planets received another bombardment five million years ago? And where is the evidence that meteorites are only five million years old?

Surely, much greater evidence than a coincidence of orbits of a relatively restricted sample of new comets is required to overturn well-grounded theories concerning the ages of craters and meteorites. Indeed, all evidence points to the existence of asteroidal, and probably cometary, objects colliding with the inner planets at least 2,000 million years ago, and it seems reasonable to assume that the minor members of the solar system known today are the survivors of this class of object rather than fragments from a recent explosion.

For many people, though, this hypothesis will undoubtedly raise the specter of that hideous monster which from time to time raises its hoary head to haunt the scientific world. I refer, of course, to catastrophism—the belief that natural phenomena can be explained by reference to sudden and violent events (such as collisions, exploding planets, and the like). Not so long ago, anyone putting forward a catastrophic theory was about as welcome in the scientific community as a communist on Wall Street, but the situation has slowly changed and a few catastrophic theories are now even considered orthodox. Thus it is no longer damnable heresy to speculate about supernovae (exploding stars) or reversals in the Earth's magnetic field as causing biological extinctions (whether it is correct or not is another matter), and the general acceptance of the big bang cosmology (surely the most catastrophic of catastrophist theories) must make cosmic catastrophies of lesser magnitude seem more respectable (especially when we see stars erupt and galaxies explode all around us!).

Nevertheless, there is still a tendency to treat as orthodox only those catastrophic events which occurred long ago or (which is the same thing in cosmology) far away. An explosion of such magnitude as envisaged by Van Flandern occurring almost next door (astronomically speaking) and only yesterday (geologically speaking) will, I feel, raise strong complaints from those committed to an "inverted anthropocentricism," to use Van Flandern's phrase.

Such philosophical considerations do not, however, constitute an argument in favor of the theory. They only point out which (equally philosophical) arguments cannot be used *against* it. Ultimately, this hypothesis (indeed *any* hypothesis) must stand or fall by the empirical evidence—and, in this instance, the empirical evidence is not encouraging.

COMET GROUPS

When the orbits of long-period comets are examined and compared, a number of unexpected coincidences arise. Thus Fujikawa's Comet (1969 VIII) has an orbit showing certain similarities to that of Comet La Hire-Bianchini-Miraldi of 1702. Now, it is possible that this latter comet is actually a short-period object (the orbit is rather poorly determined), but 1969 VIII is certainly of the long-period variety and there is no possibility that the two comets are in fact identical.

Similarly, Comet Daido-Fujikawa (1970 I) has an orbit somewhat similar to Tycho's Comet of 1577; Kohoutek's Comet (1973 VII)—not to be confused with the famous Kohoutek (1973 XII)—and Comet de Vico-Hind (1846 V) were initially suspected of being the same object until a more accurate determination of the orbit of the 1973 object precluded this possibility.

More recently, Bradfield's Comet (1979*l*) was, at first, thought to be a return of Comet 1770 II. In this case, the two comets were seen under very similar conditions, and even the absolute magnitudes were computed to be of the same value (about 7.5). Furthermore, early calculations of the orbit of 1979*l* suggested an ellipse with a period not much greater than 200 years. Nevertheless, further computations cast severe doubts upon this proposed identification, and it now seems that the similarity, physical and orbital, between these two objects is coincidental. The period of 1979*l* is currently determined to be about 306 years.

Such coincidences led to the postulation of comet groups—i.e., the hypothesis that comets moving in similar orbits were physically related, presumably being fragments of a single comet which split at some previous return. Indeed, the velocity of separation of the nuclei must have been very small in some cases, as sometimes two members of a single group have passed perihelion in a single year—Bruhns's Comet (1863 I) and Backer's Comet (1863 VI), for instance.

The phenomenon of the splitting of nuclei is a well-established one, as we have already noted, but the persistence of the subnuclei for any appreciable time after schism is not usual, and even when a secondary comet is formed in this way (e.g., the "satellite" of P/Biela), its life seems to be very limited. Furthermore, multiple schism (required to explain those comet groups with three or four members) is rather rare and, once again, the subnuclei seldom remain visible for long.

Thus it seems that we must entertain doubts as to whether this explanation of comet grouping is the correct one. In fact, the subject has been tackled anew by Dr. Fred Whipple,[18] who finds that on statistical grounds, the existence of most of the "groups" is adequately accounted for solely based on considerations of randomness.

We appear then to be faced with another of those instances where randomness gives rise to apparent patterns and groupings (something which causes no end of delight to pseudoscientists for whom such tricks of randomness provide endless "evidence" for all manner of strange and exotic speculations). It is sometimes hard to realize, but "random distribution" does not necessarily mean "even distribution." Frequently periodiclike groupings are produced by chance alone (which is not really strange when you think about it: why should chance prefer even distributions?), and it is only our willingness to see even distribution as not needing explanation and uneven groupings as "abnormal" (and therefore as needing explanation), which leads us to seek "causes" or "reasons" where none are required. (By way of aside, another example of this phenomenon, in the cometary field, may be the uneven distribution of cometary discoveries—especially of bright objects. Some years are very rich in comets whereas others are abnormally poor—all without apparent "reason.")

What may be considered empirical support for Whipple's findings has come from a consideration of the history of comets P/Tsuchinshan 1 and 2. At discovery in 1965, these two objects moved in very similar orbits and were visible simultaneously. So close were their orbits that it was postulated at the time that the comets may originally have been a single object which split during a close approach to Jupiter. In short, they seemed to form a short-period comet group, the existence of which was explained in terms of the theory of comet groups accepted at the time, namely the schism hypothesis.

More recently, however, Chinese astronomers have computed the past orbits of these two comets with considerable accuracy and have found that the present similarity (or "grouping") is merely coincidental. Rather than having been a single object in the past, the comets are simply two physically unrelated objects which just happen to be moving in similar orbits for a relatively brief period of time.[19]

Nevertheless, there does seem to be one group which involves a genuine non-chance association, the famous Sun-grazing comet group initially investigated by Kreutz. The Kreutz Sun-grazing group is both more populous and more complex than the other "groups," and has the distinction of containing the most brilliant and some of the most interesting and spectacular comets on record. There can be little doubt that this group arose from the disruption of a single large comet several thousands of years ago—a topic which we shall take up in the course of the following chapter.

REFERENCES

1. E. Everhart. "Comet Discoveries and Observational Selection." *Astronomical Journal.* Vol. 72 No. 6, Aug. 1967.

2. E. I. Kazimirchak-Polonskaya. "Review of Investigations Performed in the USSR on Close Approaches of Comets to Jupiter and the Evolution of Cometary Orbits." In *The Study of Comets (Part I)*, Donn et al (ed.). Washington, D.C.: NASA, 1976, pp. 496–536.

3. "New Light on Biela's Comet." *Sky and Telescope.* Vol. 41 No. 2, Feb. 1971, p. 84.

4. Z. Sekanina. "Fan-Shaped Coma, Orientation of Rotation Axis, and Surface Structure of a Cometary Nucleus. I. Test Model on Four Comets." *Icarus.* Vol. 37 No. 2, Feb. 1979, pp. 420–42.

5. F. L. Whipple and Z. Sekanina. "Comet Encke: Precession of the Spin Axis, Nongravitational Motion, and Sublimation." *Astronomical Journal.* Vol. 84 No. 12, Dec. 1979, pp. 1894–1909.

6. N. T. Bobrovnikoff. "On the Disintegration of Comets." *Lick Observatory Bulletin.* No. 408, 1929, p. 28.

7. F. Nölke. "Der Ursprung der Kometen und der Zodiakallichtmaterie." *Stern.* Vol. 16, 1936, p. 155.

8. R. A. Lyttleton. *The Comets and Their Origin.* London: Cambridge University Press, 1953, pp. 45, 145–49.

9. Z. Sekanina. "Relative Motions of Fragments of the Split Comets. II. Separation Velocities and Differential Decelerations for Extensively Observed Comets." *Icarus.* Vol. 33 No. 1, Jan. 1978, pp. 173–85.

10. R. A. Lyttleton. *The Comets and Their Origin,* pp. 98–99.

11. S. K. Vsekhsvyatsky. "The Periodic Comets and Their Origin." *Russian Astronomical Journal.* Vol. 25, p. 256.

12. "How Fast Do Comets Decay?" *Sky and Telescope.* Vol. 37 No. 4, April 1969, p. 221.

13. J. Bortle. "The 1976 Apparition of Periodic Comet d'Arrest." *Sky and Telescope.* Vol. 53 No. 2, Feb. 1977, pp. 152–56.

14. J. Oort. "The Structure of the Cloud of Comets Surrounding the Solar System and a Hypothesis Concerning Its Origin." *Bulletin Astronomical Insts. Netherlands.* Vol. 11 No. 408, p. 91.

15. B. Donn. *Comets, Asteroids and Meteorites.* Edited by A. H. Delsemme. Toledo, Ohio: University of Toledo Press, 1977, p. 15.

16. T. C. Van Flandern. "A Former Asteroidal Planet as the Origin of Comets." *Icarus.* Vol. 36, 1978, pp. 51–74.

17. H. B. Ridley. "Meteorites." *Journal of the British Astronomical Association.* Vol. 89 No. 3, April 1979, pp. 219–38.

18. F. L. Whipple. "The Reality of Comet Groups and Pairs." *Icarus.* Vol. 30 No. 4, April 1977, p. 736.

19. T. Kiang. "Recent Astronomical Research in China." *Sky and Telescope.* Vol. 54 No. 4, Oct. 1977, pp. 260–63.

3

Famous Comets

People sometimes ask "What was the best comet ever seen?" To this there is no positive answer. No single object can definitely be said to have wiped the rest off the field.

In 1870, Grant[1] compiled a list of comets appearing since 1066 A.D., which he considered to deserve the title of "remarkable" by virtue of their size, brilliance, and generally spectacular features. This list consisted of Halley's Comet at every return during the interval in question, plus the comets of 1106, 1264, 1402, 1556, 1577, 1618 II, 1661, 1680, 1729, 1744, 1769, 1811 I, 1823, 1843 I, 1858 VI, and 1861 II. To this list, Chambers (writing in 1909) adds 1874 III, 1881 III, and 1882 II, with consideration also being given to 1880 I, 1887 I, and 1901 I. Since the time of Chambers' writing, the list may be further extended to include 1910 I, 1910 II (Halley), 1965 VIII, 1976 VI, and probably 1927 IX. Favorable consideration for inclusion within this list may also have to be extended to 1957 III (Arend-

Roland) which, although not as spectacular as some of the others, displayed what may well have been the finest anti-tail on record, even exceeding that of Comet 1823, for which (apparently) Grant thought fit to justify this object as "remarkable."

Of course, comets are observed under different conditions from different places on the Earth's surface and recorded by people of differing temperament. What is seen as a magnificent comet by one is merely "that thing in the sky" in the eyes of another, and records left by the first will convey the impression of a far more remarkable phenomenon than those of the latter would suggest. This is especially important in records dating from ancient and medieval times when comets were regarded with intense fear. Thus the comet of 146 B.C. was said to have been "as large as the Sun" while that of 134 B.C. was said to have been brighter than the Sun. Obviously no comet in the solar system could become brighter than the Sun, and if further proof of the exaggerated nature of this report is needed, let it be noted that the same record describing its brilliance also notes that it was hidden in the Sun's rays for a time at conjunction!

Even Grant's reasoning is a little hard to follow in places. If his list is only supposed to include those comets which were remarkable because of the spectacle they provided, it is difficult to justify the inclusion of Comet 1729, as—despite its very high *intrinsic* brightness—this comet was only visible from Earth as a small tailless spot, barely visible with the naked eye. Likewise, the inclusion of Comets 1556 and 1661 but not the daylight comet of 1472 seems a curious reversal.

A somewhat better comparison may be made between comets whose orbital elements are known, enabling calculations of "absolute magnitude" to be made. In this way the intrinsically brightest known comets are (according to Vsekhsvyatsky's estimates): Halley's during its 1066 apparition (see, however, Chapter 4), 1402a, 1577, 1729, 1744, 1747, 1811 I, and 1882 II.[2] Except for the temporary outbursts of Holmes's Comet in 1892 and Humason's Comet of 1962, few objects have been more than about a fifth as bright, in *absolute* terms, as any of these. Furthermore, we have no reason to suspect, from examination of the appearances and durations of historical comets, that any comet for which an orbit has not been calculated was of noticeably higher absolute magnitude than these, and we therefore conclude that they represent something of an upper limit of observed absolute cometary magnitudes. The high brilliance of a number of other comets appears to be due primarily to their proximity to the Earth and Sun.

Of the above-mentioned comets, those of 1729 and 1747 were remote objects only faintly visible to the naked eye. That of 1811 had a head larger than the Sun and a tail some 1.3 A.U. in length, appearing from Earth as an object nearly as large as the Moon (actually, about 28 minutes of arc in diameter) with a tail which at one time reached the grand length of 70 degrees. Nevertheless, it too had a rather large perihelion distance and actually became no *brighter* than many other comets of recent centuries (Donati's of 1858 or Bennett's of 1970, for instance).

The comet of 1744, on the other hand, shone more brilliantly than Venus and displayed a tail some 90 degrees long, but unfortunately its full glory was somewhat spoiled by bright twilight. Its most remarkable feature was the system of eleven rays forming six separate tails which (after perihelion) projected in the form of a great fan from below the eastern horizon, the comet itself being too close to the Sun to be seen in the encroaching dawn. So remarkable did this phenomenon appear that the chief record (that of de Cheseaux) was long suspected by certain astronomers to refer not to the comet at all but to some freak trick of the atmosphere. It was not until corroborating reports were uncovered many years later that the description was finally accepted as being of the comet.

The comet of 1402 is regarded by many as one of the finest ever seen. According to oriental records, it had a tail "10 feet" (10 degrees?) long, and Italian records speak of it as being visible in daylight for eight days, becoming visible well before sunset and remaining in the sky until after dark. Although there are no records to the effect, if ever there has been a comet capable of casting shadows, this would probably have been it. The magnitude seems to have been around —5.

One of the intrinsically most luminous known, the comet of 1577 was also an object of small perihelion distance. It passed relatively close to Earth and became well placed as it traveled away from perihelion. Brightness estimates indicate that it would have been at least as bright as the Moon at perihelion, assuming that the rate of decrease held as far back as the time of perihelion passage. In all probability, however, this magnitude would not have been reached, as there are no records of the comet at that time. Nevertheless, it was certainly a brilliant object and was about as bright as Venus with a tail of 20–30 degrees when it became generally visible in the evening twilight.

Likewise, the Sun-grazer 1882 II (Cruls's), though passing on the other side of the Sun (except for a very short interval at perihelion), shone with great brilliance against a dark sky and is surely to be regarded as one of the most spectacular of known comets.

No discussion of intrinsically bright comets would be complete without a few words about Sarabat's Comet of 1729, already mentioned briefly in this chapter. Although often termed "the Great Comet of 1729," Sarabat's was not "great" in the usual sense, being tailless and at no time brighter than third or fourth magnitude. It was "great" not because it was spectacular (which it clearly was not), but because it was able to be seen with the naked eye at all—having a *perihelion* distance of *over 4 A.U.!*

Lyttleton[3] believed the comet to have been far in excess of all others in size, mass, and absolute magnitude. This is not strictly correct. Absolute magnitude has no *simple* relationship to cometary mass or size, and observations suggest that Sarabat's Comet had a coma diameter far less than that of (for instance) 1811 I.

Nevertheless the brightness of this comet certainly seems excessive. Watson[4] estimated an absolute magnitude of around −6, apparently based upon the fact that at a similar distance, P/Halley was some ten magnitudes fainter during its 1910 return. However, it is probable that the brightness of Sarabat's was not as dependent upon heliocentric distance as a periodic comet such as Halley, and Vsekhsvyatsky's estimate[5] of −3 is more reasonable. It is even possible that Sarabat's was so bright at these distances, not *primarily* because of its intrinsic brightness, but because it may have had a very low value of the index n. If n was equal to 2, for instance, the absolute magnitude would have been about 0 and the comet intrinsically fainter, at smaller heliocentric distances, than that of 1577.

The danger of estimating how bright a comet will become near the Sun, based upon magnitude estimates at large distances, was adequately demonstrated by Kohoutek's Comet (1973 XII), which initially was assigned a value of n equal to about 8 and a corresponding absolute magnitude of around —2, implying that it would have "overtaken" even Sarabat's if both came to an equally small perihelion distance. This, as we all know, certainly did not happen!

Personally, I rather distrust predictions of supercomets like the early (more extreme) estimates of Kohoutek and counterfactual estimates of Sarabat's. If Sarabat's was a potential supercomet, why are there no records of *other* supercomets? Absolute uniqueness is not normally the way of nature.

Thus I think that all we can really say is that Sarabat's was the brightest known comet *at a distance of over 4 A.U.*, and try to refrain from speculating about counterfactuals.

If we compare the actual intrinsic brightness of comets (rather than the absolute magnitude or how bright the comet *would* appear *if* it were located at $r = \Delta = 1$ A.U.) we can make another comparison. Actual intrinsic luminosities may be calculated by modifying the formula given in Chapter 1 in this manner

$$m\,\Delta = m_0 + 2.5n \log r$$

(where $m\,\Delta$ is the actual intrinsic magnitude and the other parameters are the same as in the original formula).

This formula shows how bright a comet becomes when the modifying effect of its geocentric distance is not taken into consideration. (The same comparison will also be possible if the geocentric distance of the comet is assumed to be 1 A.U. throughout.)

In this sense, the most brilliant comets appear to have been those of 1577 and 1882, although both these comets (especially the latter) had unusually small perihelion distances and the formula may not have held under these conditions. In fact, at Sun-grazing distances such parameters as absolute magnitude may not really matter and at least two other Sun-grazers (those of 1843 and 1965) seem to have shone as brilliantly as that of 1882.

A number of other comets, although of lesser absolute

brightness, have provided exceptional displays for their terrestrial audience, three first-class examples being 1843 I (with which we shall deal shortly), Donati's Comet 1858 VI, and Tebbutt's Comet (1861 II).

Donati's Comet was discovered as a seventh- or eighth-magnitude object on June 2, 1858. Gradually increasing in magnitude as it approached the Sun, the comet reached naked-eye visibility about August 19, and by September 2 it was of third magnitude and had a tail visible to the naked eye for some two degrees of arc. Brightness and tail length continued to increase until in early October the comet shone with a brilliance equal to the star Arcturus (0.24 magnitude) while displaying *two* appendages; a bright curving dust tail 40 degrees in length, and a straight, thin gas tail emerging from the convex side of the bright one and extending for some 41 degrees. A diffuse luminosity was noted near the head, on the concave side of the tail, and a dark region behind the nucleus resulted in the familiar "shadow of the nucleus" effect often noted in bright comets.

Observers of this great comet remarked about the spectacular "artistic" tail formation which seems to have been especially beautiful. Those who also witnessed the daylight Comet 1843 I, however, regarded that former comet as being the more spectacular, although the more southerly declination of this object probably made it less widely observed than Donati's was to be.

Three years after Donati's, an even greater spectacle was to be provided by Tebbutt's Comet (1861 II). Visible in a sextant telescope when discovered by John Tebbutt of Windsor in New South Wales on May 13, this comet displayed a tail some 40 degrees in length by June 11, at which time the nucleus was as bright as a star of second or third magnitude. At the end of the month, the comet's nucleus was as bright as Saturn (about zero magnitude), and the main tail stretched 120 degrees (!) across the sky, in addition to a series of jetlike appendages 40–50 degrees long. Observed against the dark night sky, even at midnight, the spectacle must have been superb. So large and bright was the head, and so broad the tail, that the comet was on one occasion mistaken for the rising Moon, the tail presumably being mistaken for a fan of moonlight shafts.

In late June and early July, the nucleus became visible for a few days in the daytime as a starlike point. On June 29 and 30 the daylight comet appeared rather more fuzzy than it had previously, and the daytime sky took on a peculiar yellowish hue, accompanied by a sensation of "weakness of the Sun" so strong that candles were lit indoors in some places. The peculiar sky continued into the night with the appearance of a strange luminosity, not unlike a weak auroral display. The cause of these weird effects was not hard to find—on June 29 and 30 the Earth passed through the comet's tail, at a distance of two thirds of its length from the nucleus. Undoubtedly the bright tail material surrounding the Earth gave rise to these peculiar sky effects.

Earlier I remarked that no single object stands out from the rest as obviously *the* greatest comet. While not wishing to retract this statement, one exceptional object certainly deserves special mention as, even if not standing on its own, it at least represents the extremes of beauty to which rare comets sometimes attain. Unfortunately, it appeared a long time ago and our knowledge of it must be gleaned from all-too-scanty material. I am referring to the immense comet of 1264—a truly remarkable object visible from mid-July until November of that long-ago year. Chinese records say that "it became invisible only when the Sun was high up. This lasted for over a month." If this means that it was visible until the Sun was well above the horizon, the comet's magnitude must have been, presumably at least −1 or −2 in July–August. Its tail is said to have been "over 100 feet" (degrees?) in Chinese records and to have "extended across the heavens" in both Japanese and Korean accounts, and the comet seems to have been visible for much of the night, even becoming circumpolar. Even as late as September, Chinese sources record that the tail had only "slightly decreased" in length.

SUN-GRAZING COMETS

Early in February 1843, reports began to trickle out of the southern hemisphere of a bright long-tailed comet low in the sunset sky. On February 27 it was seen at sea off Concepción (South America) in daylight as a brilliant object very close to the Sun, and on the following day thousands of people sighted a very bright dagger-shaped object up to 5 degrees of arc long, following the Sun toward the western horizon. During early March, the comet became a magnificent sight in the evening sky, displaying a tail some 60–70 degrees in length. Throughout March it provided a fine display, but early April witnessed a very rapid waning of the comet, apparently indicating exhaustion and possible disintegration. So rapid was this final fading that although the comet and several degrees of the tail were still faintly visible to the naked eye on March 30, its brightness had fallen to below magnitude 9 by mid-April, and it could not be detected at all after April 19.

Three features of this comet were especially interesting: its extreme brilliance near perihelion enabling it to become a conspicuous daylight object so close to the Sun, its extraordinarily long tail, and its remarkably small perihelion distance—only 0.005 A.U. from the center of the Sun!

The nucleus was so brilliant several hours prior to perihelion that Clark thought it sufficiently intense to remain visible in transit across the face of the Sun.[6] Though, I would think, hardly to be taken literally, this gives some impression of the intensity of the comet's light. Only on one previous occasion (on February 4, 1106) was a comet clearly recorded as shining so brilliantly near the Sun, and until the 1843 event, this early record was treated somewhat skeptically by a number of astronomers.

The tail of the 1843 comet, although (owing to perspective) not spanning a record expanse of sky, reached a length of over 2 A.U. about one month after passing perihelion. This is the longest for which we have definite records, although Messier's Comet of 1769 had an unusually long tail *before* perihelion. Most observers of this comet estimated the tail to be about 60 degrees long (which would have made it somewhat shorter, in intrinsic terms, than that of 1843), but Pingre and de la Nux claim to have traced the tail for some 90 or even 98 degrees, at which length the real extent of the tail must have been about 3.5 A.U.![7] This comet displayed remarkable tail development before perihelion but was poorly placed after passing the Sun, when the most vigorous tail growth usually occurs.

Late January and February 1880 saw another long-tailed comet of similar appearance to 1843 I, though of much lower brilliance. Orbital calculations revealed almost identical elements to 1843 I, and it looked as though this really was a return of the earlier comet, even though the orbit of the first object suggested a period of several hundreds of years and that of the second (not as well determined, since the comet faded quickly from sight) differed imperceptibly from a parabola.

It is entertaining to glance through some of the speculations, as to the comet's supposed average period, put forward in the astronomical literature of the last century. Estimates range from a fairly respectable 175 years down to the impossible 7 years stated in quite dogmatic fashion in an article appearing in the *Scientific American* for March 27, 1880. This article gave, as recorded appearances of this object, the comets of B.C. 1770, 370, 252, and 183; A.D. 33, 422, 533 I, 582, 708, 729, 882, 1077, 1106, 1208, 1313, 1362, 1382, 1402, 1454, 1491, 1511, 1528, 1668, 1689, 1702, 1843 I, and 1880 I. Needless to say, the majority of these were not even of small perihelion, let alone identical with 1843 I and 1880 I. How the alleged object of 1770 B.C. could be positively identified with the return of any modern comet is unclear (although it has been associated with P/Halley in more recent years) as our only record seems to be that of Varro, preserved by St. Augustine, telling of strange changes in the planet Venus during the reign of Ogygus (a legendary king of Boeotia in whose reign a destructive flood took place)—which sounds suspiciously like a large share of legend. Similarly, the "comet" of 1528 (described in fabulous language by Parre) is now thought to have been a great aurora and not a comet at all.

Be this as it may, the question about the period of the comet was finally settled in September 1882 with the appearance of yet another comet having similar orbital elements to the other two. At first it seemed just possible that this object might be a return of the great comet of 1880—its period having been reduced by passage through the Sun's corona—but more accurate orbital calculations soon dispelled this idea. The comets were clearly separate objects, all of the long-period variety but moving around the Sun in similar orbits. (One wonders whether the seven-year period enthusiasts would have claimed victory if 1882 II had not appeared. Seven years after 1880 I, a new long-tailed comet, 1887 I, *did* appear, moving in an orbit similar to that of the others and having a very similar appearance to 1880 I.)

The great "September Comet" (also known as Cruls's Comet) of 1882 was one of the most spectacular on record. Observed first from an Italian ship on September 1 as a bright object with a tail in the early morning sky, the comet increased in brightness as it moved toward the Sun, becoming telescopically visible in the daytime on September 14, when it was 12 degrees from the Sun, and a naked-eye daytime object with a short tail, on September 16. Just prior to perihelion (actually only about two hours before) on September 17, Elkin and Finlay at the Cape Observatory watched the comet pass onto the solar disk, vanishing at the moment of immergence like a star occulted by the Moon.

Finlay observed the phenomenon with a 6-inch (150-mm) telescope at 110 power and watched the comet move "right into the boiling at the limb (of the Sun)" at 4h 50m 58s Cape mean time.[8] Elkin, observing with another telescope, estimated the disappearance of the comet as only eight seconds earlier,[9] further commenting that the magnitude of the comet was, at the time, at least —10. The observers described the event as appearing rather similar to an occultation of a fourth-magnitude star by the full Moon. No trace of the comet could be found on the solar disk as it passed in front of it, although it is true that by the time of the transit the Sun was very low in the sky and definition was failing.

Apparently, the transit should not have occurred at this time. This is not strange—cometary ephemerides are often inaccurate—but when the comet was located again after perihelion, it was found precisely at the ephemeris position.[10] There was a similar discrepancy in one of the last pre-perihelion observations of Comet 1843 I. From a ship off Concepción, this comet was reported as being only 5 minutes of arc from the rim of the Sun when, according to orbital calculations, it should have been some 1.5 degrees distant. Could these discrepancies have been caused by a drag effect of the solar corona? Evidently not, as we now know the corona to be very tenuous, and in any case, no similar effect was noted during the perihelion passage of 1965 VIII. (Also, at least in the 1843 case, it is doubtful if the recorded position was accurate—how could the comet be a conspicuous naked-eye object, to someone who did not know of its existence, only 5 minutes of arc from the Sun?)

After perihelion, 1882 II was again visible in daylight and became increasingly prominent in the predawn sky with a brilliant multiple tail at least 30 degrees long. Just how bright it became is uncertain—direct application of the brightness formula gives about −17 at perihelion, but (as observations were apparently lacking) we cannot be sure of this, although the brightness must have been somewhat of that order. Even ten days after perihelion, the comet and 10 degrees of its tail were visible when first-magnitude stars had disappeared in morning twilight; and on November 12, the intensity of the tail near the head was said to correspond to extrafocal images of third-magnitude stars.

The famous division of the nucleus was first noted on September 30, and in early October at least four separate nuclei were visible—one bright nucleus, a second considerably dimmer, and two faint ones, with a fifth very faint (and probably transitory) one suspected on one or two occasions. The two main nuclei apparently acted as separate though "overlapping" comets, each with its own tail. This gave a peculiar effect, especially during November when the two tails diverged in a formation shaped like the Greek letter gamma (γ).[11]

Another peculiar phenomenon observed during October was the series of from six to eight (although Barnard apparently placed the number as fifteen or higher) temporary satellite comets, receding from the main object at a rate of approximately 1 degree per day. Orbital elements were even calculated for one of these objects, seen for several days by Schmidt, and these revealed that a perihelion passage of 0.0184 A.U. took place on September 24.2,* although such values must be considered as being rather hypothetical under the circumstances.

Apparently these minor comets were fragments torn away from the main mass by solar gravity, separating from it at maybe five times the velocity of separation experienced by the subnuclei within the head.

At this time the comet also displayed an anti-tail, appearing at

* The time for most cometary events is recorded to the decimal of the day.

times like a sheath of light surrounding the comet and projecting at least 3 degrees in the direction of the Sun. A drawing by Bredikhin on October 17 showed the comet to possess a triple main tail, composed of type I, II, and III components.

Even as late as mid-January 1883, the comet was an easy naked-eye object of magnitude 3 or 4 and with 15 degrees of tail still showing. At this time, the comet was over 2.5 A.U. from the Sun and nearly 2 A.U. from the Earth. Tebbutt of Windsor could still detect the comet with the naked eye as late as February 13, but apparently it faded more rapidly during March and April and was reported as being very faint and diffuse by late April—fainter, it would seem, than its earlier brilliance would have led one to expect. This may be indicative of a certain degree of exhaustion, although the comet did not fade out very rapidly in the manner of the other Sun-grazers and was still visible as late as June 1, on which date it was last measured by Thome at Córdoba.

Another Sun-grazer made a brief but impressive appearance in 1887, after which no more appeared until December 11, 1945, when Du Toit discovered a seventh-magnitude comet moving through the constellation of Triangulum Austrinus. Although only briefly observed, orbital calculations showed the object to be moving in a path having very close similarity to that of 1882 II; it was, therefore, yet another member of the Sun-grazing group.

On the morning of September 14, 1963, Z. M. Pereyra of Córdoba discovered a bright (second-magnitude, according to initial estimates) comet with a long tail, very low in the morning twilight. This proved to be another Sun-grazer, with an orbit very similar to those of 1843 I and 1880 I; its perihelion passage had already occurred on August 24. The comet displayed a tail 12 degrees long at first, but faded rapidly as it moved away from the Sun and Earth, possibly undergoing a division of the nucleus in November when the last observations were made.

Interest in the Sun-grazing group was rekindled by this comet and, more especially, by that of 1965, the only Sun-grazer of the present century to live up to the grand reputation of its nineteenth-century predecessors. (This comet will be discussed in a later section of the present chapter.)

Since 1965, one further Sun-grazer has appeared—Comet White-Ortiz-Bolelli (1970 VI). First observed by Wollongong

University student G. White as a bright (at least first-magnitude) object 12 degrees from the Sun in the evening twilight, the comet provided a brief naked-eye spectacle with some 15 degrees of tail visible at times. However, like 1963 V, it passed on the far side of the Sun and faded very rapidly.

All these objects (with the possible exception of 1887 I, whose orbit was not very well determined) were almost certainly members of a single comet group, presumably originating with the disruption of a single comet near perihelion—but more on this later.

In addition to the above-mentioned known or strongly suspected members of this group, a number of "possibles" may also need inclusion. The most likely of these is the comet of 1668, a bright long-tailed object studied by Gottignies. Not so frequently included among group members are the bright comets of 1689 and 1695, although it is quite possible that these objects really should be included as members. These three comets, according to the most widely accepted orbital determinations, did not pass as close to the Sun as the later Sun-grazers, although their perihelia still lay within 0.1 A.U.

Likewise, Comet 1702a is strongly regarded as another suspect, although observations of this object are so few and so poor that orbital calculations have not been made. Nevertheless, the three tolerably useful observations (obtained on the evenings of February 27, 28, and March 2) lie very close to those expected for an object moving in the orbit of 1882 II.

During the solar eclipse of May 17, 1882, a small but very bright comet with a strongly curving tail was observed almost within the corona. It was not observed before or after the eclipse, and therefore no orbit could be calculated, although it has been noted that the comet's position was within 0.1 degrees of a comet moving in the orbit of 1843 I, if five hours prior to perihelion.[12]

Some other "possibles" have been noted from time to time. For instance, a few months after 1843 I appeared, a large sunspot was noted (during a period of minimal sunspot activity), and the suggestion was raised at the time that this phenomenon might have been caused by a large meteoritic body moving in an orbit similar to that of the comet and striking the Sun. This explanation is not very likely, but is included here for what it is worth.

Another "not-very-likely" is the object of a single report from Ceylon, dated 1666. Although generally this was believed to be a mistaken reference to the comet of 1668, Lynn[13] suggests that 1666 might be correct and that this is a reference to a further member of the group.

Speculation has also surrounded the daylight "star" seen by several people at Broughty Ferry (Scotland) at 11 A.M., December 21, 1882.[14] This object was said to have been of a milky white appearance and to have appeared crescent-shaped "through the glass." Nevertheless, there was no mention of a tail (and Sun-grazers near perihelion usually have brilliant tails) and a Sun-grazer coming to perihelion on that date should have been discoverable from the southern hemisphere before and after perihelion (although a very faint object which flared at perihelion and subsequently disintegrated very rapidly might conceivably account for the Scottish observations). Furthermore, the Broughty Ferry observers also mentioned that the object was above the Sun's path, which would seem to indicate a position north (probably several degrees north, as it could not have been *very* close to the Sun) of the ecliptic, which is not suggestive of a Sun-grazer.

Other reported or suspected objects seem even less likely.

For instance, the comet allegedly found by Klinkerfues while searching for P/Biela in 1872 does not appear to have been a Sun-grazer, even though its perihelion distance was probably quite small. Likewise, the mysterious daylight object of 1921 seems not to have been a member of the group, and we must also exclude the brilliant starlike object (as bright as Venus) observed for only ten minutes after sunset on June 26, 1915, by Anna Caroline Brooks and her father (the famous comet hunter W. R. Brooks) as well as by several of their companions.[15]

Comets seen during total eclipses of the Sun should always be investigated as possible Sun-grazers, although that of 1893 and the suspected eclipse comet of 1963 do not seem to have been associated. In fact, of the four eclipse comets of the past century, only that of May 1882 emerges as a likely Sun-grazer.[16]

The appearance of Comet Ikeya-Seki (1965 VIII) aroused interest in the Sun-grazing comets, and recent investigations by B. G.

Marsden have added new and interesting information to our store of knowledge about these objects.[17]

Marsden finds, for instance, that the orbits fall into two distinct subgroups in which the lines of nodes are separated by about 20 degrees. Subgroup I consists of the comets of 1843, 1880, and 1963; and also those of 1668, 1695, 1882 (May), and 1887—if, indeed, these latter four are bona fide members of the Sun-grazing group. Subgroup II consists of the comets of 1882 (September), 1945, 1965, 1970, and (possibly) 1689 and 1702 if these are included. The most accurately determined orbits are those of the comets 1843 I and 1963 V in subgroup I and of 1882 II and 1965 VIII in subgroup II. Backward computations of the orbits of these objects may, therefore, provide information as to the origins of these subgroups and, maybe, even the origin of the Sun-grazing group itself.

Marsden found, by backward calculation of their orbits, that the comets of 1882 and 1965 were almost certainly a single object at the previous return, and he found also that this return must have occurred early in the twelfth century. Now, on February 4 or 5 in the year 1106, a bright star (some records say a comet) was seen from Europe and reported as "one foot and a half" (presumably about 1.5 or 2 degrees of arc) from the Sun during the daytime, and on February 7 a comet was seen after sunset from Palestine and Constantinople. Beginning on February 9, a bright comet with a tail up to 100 "feet" long was recorded in China, Japan, and Korea.

This certainly sounds like the required comet, but there are difficulties. For instance, European records imply that it moved northward, which would preclude any association with the Sun-grazing group. However, oriental records do not mention any such northerly motion and seem quite consistent with the Sun-grazing nature of the object. Could the northern observations have been due to an aurora—a possibility suggested by Marsden? It is not really difficult to imagine how such an error could have arisen. In the days when the mere mention of a comet was enough to cause palpitations in even the bravest heart, people were apt to mistake even odd-shaped clouds illuminated by the setting Sun for comets (as is said to have actually happened in the middle ages and even later), and once a rumor began to circulate that a bright comet had been observed, any odd phenomena could easily be mistaken for it, especially in view of the strong southerly declination of a Sun-grazing comet, possibly making an auroral display more spectacular for a northern observer. (Incidentally, unless we imagine that such a mistake is a rather quaint feature of a simpler age, it should be remembered that Jupiter was mistaken for Kohoutek's Comet [1973 XII] in 1973. Times and people do not *really* change.)

Similar calculations of the orbits of 1843 I and 1963 V also show these objects to have been closer at their previous returns (about the end of the eleventh century), although in this case it would seem that they were still separate objects. Presumably, the schism occurred on the previous return in the fourth century.

It is possible that one of these objects was observed as the comet of 1075, and we may speculate that the other might have been the "Blazing Starre seen near unto the Sonne" on Palm Sunday (April 19), 1077—if indeed this latter object was a comet and not the planet Venus at inferior conjunction, as suggested by Pingre. A Sun-grazer appearing at that time of year would have had a fairly strong southerly declination and might have escaped observers in the northern hemisphere, except for a few hours in daylight near perihelion.

Tracing the orbits still farther back into time, we presume that the comets appeared sometime during the fourth century A.D. Furthermore, as fragments of split comets do not normally last long, we may surmise that there were only two Sun-grazers appearing at that apparition, the subgroups themselves forming with the breakup of these two objects at the fourth- and eleventh-century returns.

There appears to have been at least one daylight comet in the decade 360–70. Ammianus Marcellinus writes that during the reign of Jovian, "comets are said to have been visible in daytime." Emperor Jovian only reigned for seven months from the latter part of 363 until early 364, indicating that the comet probably appeared late in 363. As a matter of fact, a comet did appear in August and September of that year, as recorded in the Orient and elsewhere, but its position in the sky (and even its visibility at that time of year) is not consistent with its having been a Sun-grazer. Whether this is the same comet as the daylight object is not, however, clear; and neither is Marcellinus' statement that "comets" were seen in the daytime. Are we to believe that there was more than one daytime comet in that seven-month period? (Not very likely, but we cannot be sure.)

The year 368 has also been given as a year in which a daytime comet appeared, although I have not personally been able to find reference to such an appearance. Perhaps it was only the 363 comet misplaced, but it is also possible that a Sun-grazer appeared in that year—the evidence is simply too scant.

A daylight comet certainly appeared in the year 302 and was recorded by observers in the Orient in May or June of that year. This time of year is actually a very bad one for observing Sun-grazers; actually from late May until early September the comets would remain too close to the Sun in the sky to be visible, unless they reached daylight visibility for a very brief time around perihelion. Could the lack of information about the nighttime appearance of the comet of 302 imply that it was *only* visible in daylight? This could be seen as evidence (albeit not compelling evidence) that this comet was a member of the Sun-grazing group.

The return before this looks a little more promising—at least for the subgroup II progenitor. In the year 371 B.C. (approximately), a very large comet was observed by the Greeks, and the records were sufficient for the cometologist Pingre to calculate some *very rough* elements of its orbit. Needless to say, the elements cannot be considered sufficiently accurate or sufficiently comprehensive to satisfy the orbital computers of today, but they are of interest as they reveal a certain similarity to those of Comet 1882 II and the subgroup II Sun-grazers. Of course, Pingre calculated these elements long before any members of the Sun-grazing group were recognized—1882 II still lay far in the future—and there can be no question of "reading out" Sun-grazing elements due to preconceived notions.

Most of our information concerning this early object comes from Aristotle, who includes it among the four comets which he discusses, differentiating it from the others (on each occasion in which he refers to it) by a word translated in English as "great."

None of the other three comets (one of which may have been a return of P/Halley) apparently deserved that title.

Aristotle reports that the comet first became visible after sunset about the time of the winter solstice in very clear and frosty weather. The tail, he reports, stretched across the sky "like a great ribbon" (or, according to some translations, "like a great leap"—the word for "leap" and "ribbon" being very similar, although "ribbon" is certainly the most descriptive in this instance). From his descriptions, we can estimate the length of the tail as being about 60 degrees.

Presumably, the progenitors of each subgroup were themselves originally part of a single comet which split, according to Dr. Marsden's estimates, sometime between ten and twenty revolutions ago.

Incidentally, it is not necessary to picture this "original Sun-grazer" as an object of superbrilliance. Observations of other cometary schisms (e.g., those of P/Brooks 2 and P/Biela) reveal the secondary fragments to be frequently as bright, or brighter, than the parent comet; and there is no real indication that the comets of 1106 A.D. and 371 B.C. were brighter than the recent Sun-grazers. In fact 1882 II remained a naked-eye object for longer than the comet of 1106 (at least, this is indicated by the records), and this may be indicative of greater intrinsic brightness. In general, I do not think that the original Sun-grazer was necessarily any brighter than those of recent centuries, and the oft-stated supposition that it must have been some sort of supercomet is, I believe, based upon a misunderstanding of confusing mass with absolute brightness.

We may, therefore, dare to speculate about the development of the Sun-grazing group somewhat as follows: Some unspecified number of millennia ago, a large new comet became deflected (presumably by stellar perturbations) into the central solar system on a near-collision course with the Sun, and it was subsequently captured in an elliptical orbit having a relatively short period of about 1,000 years. Either on this or, more likely I think, on some subsequent passage through perihelion, the nucleus divided into two approximately equal fragments which persisted as separate comets for some ten to twenty revolutions, having their orbits gradually altered by planetary perturbations, until first one (probably in the fourth century A.D.) and then the other (probably in the year 1106) divided into at least three fragments each, which have returned separately during the last few centuries as members of the Sun-grazing group.[18]

Separation at similar velocities to those observed in the breakups of 1882 II and 1965 VIII could account for comets appearing about eighty or one hundred years apart at the following return (assuming a period of between 800 and 1,000 years). For fragments to return in relatively quick succession (as, for instance, the three subgroup II objects 1945 VII, 1965 VIII, and 1970 VI), a less violent separation would be required, such as that of the division of nucleus B of 1965 VIII which was seen as a very close triplet in early November of that year.

We may well be witnessing the final demise of the Sun-grazers. Of all the observed objects belonging to this group, only 1882 II did not *rapidly* exhaust after perihelion, and the possibility remains that one or (less probably) two of its subnuclei may be visible at the next return—although even this is by no means certain, and some signs of exhaustion *may* have been apparent in the late fading of this comet. If indeed we have seen the end of the Sun-grazing group, we must count ourselves fortunate that the original schism did not occur one or two revolutions earlier as we may then have never known of the existence of these fascinating and spectacular comets; and since most of our knowledge concerning the metallic contents of comets has come from the near-perihelic observations of 1882 II and 1965 VIII, our knowledge of the constitution of comets as well as their behavior at very small heliocentric distances would, accordingly, have been greatly impoverished.

As already mentioned, several small "satellite" comets were observed near Comet 1882 II soon after perihelion. These were apparently very tiny fragments and did not persist for more than a few days, but it is not impossible that similar satellites, separated at an earlier return, may have persisted and returned as separate comets at some later epoch. Is it possible that other comets of very small perihelion distances, but moving in orbits otherwise dissimilar to Sun-grazers, originated in this way? In particular Comet 1887 I (for which the Sun-grazing orbit is not entirely satisfactory, despite its very small perihelion distance) may need to be explained in this manner rather than as the result of a simple low-velocity schism.

Another possibility is Kirch's Comet of 1680 which, despite a perihelion distance of only 0.00622 A.U., has an orbit dissimilar to those of other Sun-grazers in all other respects. Nevertheless, as pointed out by Kreutz (one of the first people to study the Sun-grazing group in detail), the orbit of this comet passes within 0.0005 A.U. of that of 1882 II—a strange coincidence, if coincidence it really is. In this respect Marsden's "wild speculation," regarding the alleged split of the great comet of 371 B.C. as possibly being the separation of the 1882 and 1680 comets, may not be so wild after all. (This phenomenon, if it really occurred, must have been more akin to the separation of satellite comets from 1882 II than to the low-velocity splitting of its nucleus.)

There may also be some relevance in the appearance of comets having very small perihelia during times when Sun-grazers are prevalent. For instance, between 1843 and 1882, comets with perihelia of less than 0.1 A.U.—but otherwise apparently unrelated to the Sun-grazing group—appeared in the years 1847, 1865, 1872 (possibly), 1874, and June 1882. There was a dearth of such comets until 1931 (and also of true Sun-grazers), but others were seen in 1954, 1961, 1962, and 1970.

At first sight it appears especially remarkable—considering the paucity of comets having perihelia within 0.1 A.U. of the Sun—that the Sun-grazer of 1963 should have been heralded by two such small-perihelion objects (in 1961 and 1962) appearing only nine months apart. Likewise, it seems a strange coincidence that the Sun-grazer of 1970 should have been preceded only three months by a comet passing a mere 0.07 A.U. from the Sun.

Nevertheless we must beware of jumping to premature conclusions. There is really no concrete evidence to suggest that these comets were in any way associated with the Sun-grazing group, or that these apparent clusterings were anything other than pure chance. In fact one of these comets, Daido-Fujikawa (1970 I), has an orbit showing greater similarity with that of the great daylight comet of 1577 than with any of the Sun-grazers (any physical relationship between these two comets is not implied by this observation).

Irrespective of the relation of other small-perihelion comets, the preceding account of the separation of group members themselves seems to be the most economical explanation of the Sun-grazing phenomenon. Other suggestions, however, have been put forward at various times, the most bizarre (at least according to our present understanding of stellar distances) being one made last century and postulating the expulsion of these comets from the star Sirius—a suggestion inspired by the fact that the star lies near the group's aphelion point on the celestial sphere (8.7 light years or 5,501,880 A.U. distant from the Sun!).

Other suggestions concerned the possibility that these comets might originally have emerged from solar prominences, a hypothesis which at least avoids the distance problem of the latter, but about which little more needs to be said.

More modern—and more credible—theories include the postulation by Z. Sekanina of a collision between two protocomets at a heliocentric distance of about 1 A.U. Such collisions would be extremely rare (but then so are Sun-grazers!), and the process is certainly more complicated, and therefore less likely to have arisen in nature, than the one discussed above.

GREAT COMETS OF THE TWENTIETH CENTURY

Great comets provide us with one of nature's most beautiful sights. Seen against a clear dark, star-studded sky, with their fuzzy heads and long ghostly tails streaming away like plumes of luminous smoke floating in some unseen cosmic breeze, they leave an indelible impression on the memory of even the most hard-headed among us.

Like similar unofficial classifications (such as shooting stars and fireballs), there is no hard and fast dividing line between large "ordinary" comets and "great" comets—just how large a comet has to become before being eligible to qualify as a member of this club of the cometary elite is not a straightforward matter. High brilliance alone does not make a comet "great"—nor does its absence preclude greatness. For instance, comets 1880 I and 1887 I are both considered to have been great comets even though neither was exceptionally bright. Even the famous great comet of 1680 was not of unusually high luminosity. Conversely, the magnitude of Giacobini's Comet (1906 I), in no sense a great comet, actually rivalled that of the 1910 return of Halley's for a very brief period, whereas Seki-Lines (1962 III) outshone them both!

Nevertheless, if we assume that a comet of sufficient brilliance to be conspicuous to the man in the street, one that has a bright naked-eye tail of at least 10 degrees, is justifiably termed "great," we may list as twentieth-century great comets, the following: 1901 I, 1910 I, 1910 II, 1927 IX, 1947 XII, 1948 XI, 1957 III, 1965 VIII, 1970 II, and 1976 VI.

Counterfactuals should, perhaps, be avoided, yet we may dare to mention a number of "great comets if . . ." that have raised their disappointing heads among the catalogue of twentieth-century luminaries. Thus 1914 V and 1962 VIII would probably have been great . . . if they had come nearer the Sun, 1963 V and 1970 VI would likewise have been great . . . if they had been more favorably placed as seen from Earth (which in practice would have meant the former coming to perihelion two or three months later and the latter two months earlier). Other possibilities include 1957 V, which would have had a much longer tail had perspective been more favorable, and several objects of high intrinsic brightness but large perihelion distance, among which we may include Meier's Comet 1978 XXI.

If all these objects had fulfilled their potential, the list of twentieth-century "greats" would have been impressive indeed. Even so, the list reveals this century to have been quite normal cometwise and not as barren as is sometimes stated.

It is true, though, that this century has not brought forth a great comet of high intrinsic brightness—in the class of, say, those of 1811 and 1882—and it is comets such as these which are likely to remain visible for long periods (being still bright when relatively far from the Sun) and which attract wide attention from the general public.

Also, the majority of this century's great comets have had an unhappy tendency to be early morning objects, and most have been south of the equator. As most of the world's population lives in the northern hemisphere (often under city lights) and does not make a habit of rising hours before the Sun, it is not surprising that many people alive today have not seen the beautiful spectacle of a great comet.

Of course, there have been other comets clearly observable with the naked eye, among the more prominent being Daniel's (1907 IV), Beljawsky's (1911 IV), Brooks's (1911 V), Delavan's (1914 V), Finsler's (1937 V), Jurlof-Achmarof-Hassel (1939 III), Cunningham's (1941 I), and De Kock-Paraskevopoulus (1941 IV). (Incidentally, mention of Brooks's Comet [1911 V] brings to mind an error often found in elementary astronomy books—namely, the designation of this object as P/Brooks 2. This periodic comet did, indeed, return to perihelion during 1911 and was designated 1911 I, although the return was very unfavorable and the comet was only seen on a few occasions in late 1910, when its brightness was about magnitude 16 and no tail was visible. On the other hand 1911 V, Brooks's last official discovery, became a bright naked-eye object of the second magnitude and at one time had a tail some 20 degrees long.)

These and other comets (some of the more recent we will meet shortly) have proven to be interesting naked-eye objects, although none of them could really be termed "great." To those which could be thus termed, we shall now turn.

Viscara's Comet (1901 I)

Viscara's Comet, the Great Comet of 1901, was first located on April 12 as an object of magnitude 1 or 2 and with a definite tail, low in the morning twilight. It moved toward perihelion (April 24), on which date it was brighter than Sirius and telescopically visible for some fifteen minutes after sunrise.

After passing perihelion, it became an evening object, best observed from southern latitudes, and in mid-May it displayed a main tail 15 degrees long, a fainter secondary tail 45 degrees long, and two faint, shorter appendages between the primary and secondary tails. On May 15, the nucleus was found to have divided into two components, one about a magnitude brighter than the other.

The comet remained a naked-eye object until about May 23 and was subsequently traced telescopically until June 14, by which date its magnitude had decreased to around 10 or 11.

1910 I

As the world awaited the arrival of Halley's Comet in 1910, another large object approached the Earth and Sun, unseen in the bright morning twilight, until spotted by miners in the Transvaal on January 12 and mistaken for Halley's Comet (which was still too faint to be observed with the naked eye). The comet was again detected by railway workers at Copier Junction (Orange Free State) on January 15 and by Innes (Johannesburg) on January 17, after he had read the newspaper report.

The world at large first heard the news by means of telegram from Johannesburg telling of the discovery of a "great comet"; however, the word "great" somehow became garbled in transmission, and what was in fact announced was the discovery of "Drake's Comet!" The object, accordingly, became known by this name until the mistake was found and rectified; and it is now popularly known as "the Great January Comet," "the Miner's Comet," "the Johannesburg Comet," or (more frequently) "the Daylight Comet."

This last popular appellation arose because of the comet's extreme brilliance at perihelion, rendering it a naked-eye object with a 1-degree tail, only 4 degrees from the Sun in the clear daytime skies of South Africa and observable elsewhere as a very bright spot near the Sun, with a varying length of tail.

Like Viscara's Comet nine years earlier, Comet 1910 I crossed into the evening sky after perihelion where it became, toward the end of January, a beautiful sight with a brilliant broad, curving tail (at times blending with the zodiacal light and giving the impression that the comet was associated with a large nebulous glow) and a straight type I secondary. Jets and an anti-tail were also reported. Fortunately for northern observers, the comet moved northward after perihelion and became a prominent object after sunset for European latitudes. Maybe this was a mixed blessing as, in addition to providing European observers with the chance to view a spectacular comet, the object also aroused a good deal of superstitious terror, especially in Portugal where peasants are said to have flocked in thousands to the seashore, crossing themselves in fear as they watched the comet emerge from the darkening sky.

One observer in the south of England, viewing the comet far from the interfering glow of city lights, was able to trace the tail for some 45 degrees in late January and remarked that this was the largest comet he had observed since Donati's of 1858, all intervening objects being only "poor little things" by comparison.

Moving away from the Sun and Earth, the comet faded slowly—to about magnitude 7 in mid-February and 9 by March. The last observation was reported on July 9, when the comet was a fourteenth-magnitude object, over 3 A.U. from the Sun.

Comet Skjellerup-Maristany (Skjellerup's Comet) (1927 IX)

Late November 1927 witnessed the arrival of yet another bright comet. First observed by O'Connell on November 28, it was also sighted independently by Skjellerup and a number of other people (including one lady who discovered it from her bed) in early December as a bright twilight object with a tail. Because of its high celestial latitude, it could be observed from the southern hemisphere both after sunset and before sunrise during the early period of its visibility.

As it approached perihelion (December 17), the comet became increasingly bright and was visible in daylight by December 15, when a photograph was obtained from Lowell Observatory at midday with the aid of an infrared filter—probably the first daytime cometary photograph. On December 15, 16, and 17, the comet was a naked-eye object throughout the daytime hours even when it was little more than 1 degree from the Sun. Glowing with the intense yellow light of sodium, it certainly outshone Venus at maximum brilliance—some estimates even put the brightness as high as −10, although Baldet's estimate of −6 is probably more realistic. In any case, it does appear to have been a brighter object than either 1901 I or 1910 I, and it is a pity that it was so poorly placed in the bright twilight sky.

Still a daytime object on December 20, the comet faded rapidly thereafter although a 35-degree tail was seen in the morning twilight on December 29, persisting until about January 3. The last observation of the comet was made at Johannesburg on April 28, by which time its magnitude had decreased to 13 or 14.

1947 XII

Seen by many observers in the bright twilight of December 8, this object is reported as having possessed an orange head (presumably betraying the presence of sodium emission) and a tail some 25 degrees in length. The general appearance of the comet was said to have resembled that of 1910 I. It must have been very bright as various descriptions remarked that it was "brighter than Venus" and one magnitude estimate even puts it as high as —5, although there do not appear to have been any daytime sightings and these values may be a little high.

On December 10 the nucleus was seen to have divided, and the tail itself became multiple some six days later. On December 17, two main tails were visible, flanked by two shorter appendages and a fifth narrow one, all spread out in an impressive fan.

By Christmas, the comet's magnitude was estimated to be 6.2 by Wood at Sydney, and it had fallen to magnitude 10.5 by January 20, as estimated by Van Biesbroeck at Yerkes Observatory.

1948 XI

Almost one year after the discovery of the above, on November 1, 1948, an eclipse of the Sun was seen from Nairobi; and during totality, a comet appeared only 2 degrees from the Sun, rivalling Venus in brightness and with a long tail lagging far behind the radius vector and pointing toward the horizon.

The comet vanished with the return of sunlight; then from November 6 it again became visible in the predawn twilight as a bright object with a 30-degree tail, and on November 10 it was photographed from Palomar Mountain as a second-magnitude luminary with some 15 degrees of tail still observable. One observer saw the tail emerging from behind the dawn-lit horizon and at first mistook it for a shaft of sunlight until the head appeared and put an end to all dispute.

The comet remained a naked-eye object until about December 10, but had faded to approximately magnitude 16 by April 2, when the last observation was obtained.

Comet Arend-Roland (1957 III)

On November 6, 1956, a comet of magnitude 10 was discovered photographically by Arend and Roland of Uccle during the course of a minor planet program. Early orbital determinations indicated that the comet was still far from perihelion and should become a bright naked-eye object during April and May of the following year.

As it moved toward the Sun, the new comet displayed a number of brightness fluctuations of up to two magnitudes, a feature which had also been noted for P/Halley in 1909, when at similar heliocentric distances. It seems that discovery may have taken place during one of these fluctuations, as magnitude estimates later in November put the comet at about magnitude 12.

Nevertheless, the comet brightened well and by late January 1957 was about eighth magnitude with a tail over 1 degree long. In early April, a tail of up to 5 degrees had become visible and the total brightness may have been as high as second magnitude.

The best spectacle came, however, during the last two weeks of April when the comet emerged from the evening twilight and became well placed for northern observers. At this time, the magnitude was estimated to be about zero and the main tail some 30 degrees in length, but the true glory of the comet was the spectacular anti-tail, appearing as a long bright spike—apparently directed sunward—which achieved greatest prominence between April 22 and 24, being then some 15 degrees in length. Thus the whole comet stretched over 45 degrees of the sky at this time and was described by English astronomers as having been the largest and most spectacular since 1910 I.

After April 28, the anomalous tail started to fade away and the comet itself grew steadily fainter as it moved away from the Sun and Earth. In June, the main tail was very faint and less than one degree long, the comet itself being little brighter than ninth magnitude. Nevertheless, the comet was followed with large telescopes for more than a year after perihelion, the last observation being by Dr. Elizabeth Roemer at Flagstaff on April 11, 1958, when the comet was of magnitude 21.

Since Arend-Roland, a number of bright and interesting comets have been observed, three of which must surely be termed "great" and will be discussed at greater length a little later.

Nevertheless, some of the others need a few words as well, since even if they were not "great" comets, they were certainly bright and must have contributed to the general revival of interest in comets which has been gathering momentum in recent years.

The first such object appeared even before Arend-Roland was out of range—Mrkos's Comet (1957 V), discovered when near perihelion at the beginning of August of that year at almost first magnitude (see Plate 10). Actually, it was intrinsically about as bright as Arend-Roland, and the tail (in real terms) was almost as long, but the dust tail was not as intense as that of the earlier comet, and due to the Earth-comet geometry, it did not span more than 5 degrees of the sky.

The next comet of equal interest was Seki-Lines (1962 III), which became a brilliant object low in the southwestern sky in late March of that year. As observed at that time by Mr. Alan Marks of Guildford (near Sydney), the comet had a dense and brilliant nucleus and a fan-shaped tail—plainly visible with the naked eye even in bright twilight—taking the form of a hollow cone with two brilliant side streamers and a comparatively dark central region. Mr. Marks estimated the nucleus to be "as bright as the planet Mercury," and "solid as a planetary disk." The final pre-perihelic observation was by Mr. Marks on the evening of March 27, when the comet's magnitude was —1.

Actually, why it was not seen nearer to perihelion is something of a mystery. Perihelion occurred on April 1, and the comet was then located 2 degrees from the Sun and (if the rate of brightening displayed in the preceding weeks continued right up to perihelion itself) as bright as magnitude —7.5. Normally, a comet this bright should blaze out conspicuously in broad daylight rather in the manner of the daytime objects of 1910 and 1927, yet no such reports are known for Comet Seki-Lines. In fact, from Mr. Marks's observation, until Mr. K. Hindley's observation just after sunset on April 3 (when the comet was estimated as magnitude —2.5), there is a complete and unexpected lack of sightings. (This is, perhaps, not without some irony—after all,

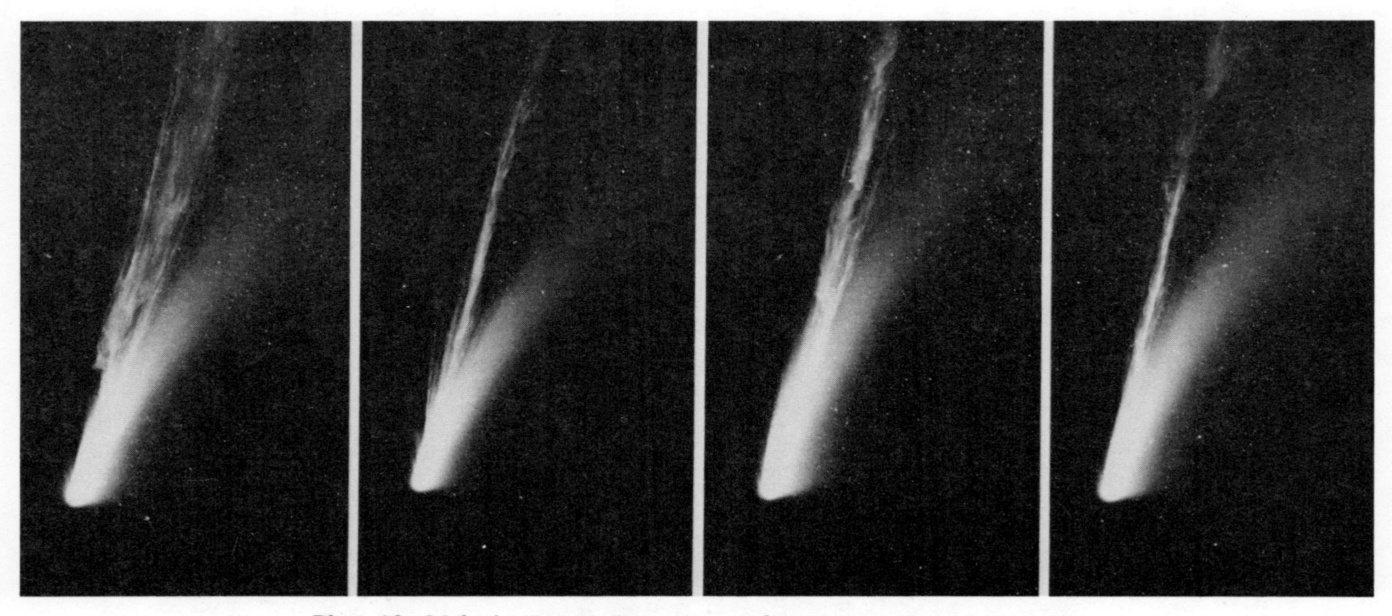

Plate 10. *Mrkos's Comet, four views taken August 22, 24, 26, and 27, 1957.*

the comet *did* come to perihelion on April Fool's Day!!) Clearly, the predicted perihelic brightness was not reached.

After perihelion, the comet became a spectacular object with a bright dust tail about 15 degrees in length and a short anti-tail visible for awhile. In fact, this comet could almost qualify for the title "great" and was certainly one of the major objects of the 1960s.

One year later, Ikeya's Comet (1963 I) reached approximately second magnitude and displayed several degrees of tail to the naked eye, but was not especially conspicuous.

The next major comet was a great one—Ikeya-Seki (1965 VIII), which we will discuss more fully later—but there were no other naked-eye objects during 1965 or 1966, although Rudnicki's Comet (1967 II) became visible in a tiny 2.5-power opera glass in late 1966, but it had little tail and was not really impressive.

Comet Mitchell-Jones-Gerber (1967 VII) was almost certainly a very bright object when near perihelion in mid-June of that year, but its elongation from the Sun was very small and it was not discovered until early July, by which time it had faded to magnitude 4 or 5. Nevertheless, the head could still be seen with the naked eye as a small fuzzy "star," and a magnificent 7-degree tail was grandly displayed through a pair of binoculars.

The next comet of considerable brightness was Tago-Sato-Kosaka (1969 IX), visible with the naked eye in late December 1969 and early January 1970 as an object of approximately fourth magnitude with an impressive tail. In late December, the comet could be seen in the southern sky for southern hemisphere observers, and although inconspicuous with the unaided eye, it was a magnificent sight in binoculars; the tail attained a quite high intensity by the end of the year. Although the intensity dropped somewhat in early January, the length of the tail increased and became more conspicuous with the naked eye. The naked-eye tail attained a maximum angular length of some 12 degrees on January 6 and 7, corresponding to a real length of over 19 million kilometers.

TWO GREAT COMETS, A THIRD THAT WASN'T, AND A FOURTH THAT SURPRISED US ALL!

Ikeya-Seki (1965 VIII)

In October 1965, a comet appeared which outshone any other seen during the present century and equalled in brilliance the very brightest recorded comets of the past. Unfortunately, its show was rather brief and confined to the very early morning hours where few people would see it unless special effort was made. Moreover, it was predominantly a southern object, and its arrival coincided with a good deal of cloudy weather in many parts of the globe.

Early in the morning of September 18, 1965, in the dark predawn skies of Japan, swept clean by a typhoon some days earlier, two young amateur astronomers—K. Ikeya and T. Seki—almost simultaneously (though independently) noticed a small fuzzy blob of light "shining," in Seki's words, "like a streetlight on a foggy night." Little did they know that in five weeks this "foggy light" would turn into one of the most brilliant comets on record!

At discovery it was of magnitude 7 or 8 and only 1 minute of arc in diameter. It was slightly elongated away from the Sun, but did not display any visible tail. Moving toward the Sun, it rapidly brightened, however; and by September 26 some 1.5 degrees of tail were visible on a photograph taken at Table Mountain. At the month's end, it had crossed the threshold of naked-eye visibility, and about 20 minutes of tail had become visible in small telescopes. Nothing, however, pointed to the spectacle to come—not at least until a preliminary orbit had been calculated, by early October, and the similarity between this and the orbit of the brilliant Comet 1882 II noted.

Clearly, the new object was a "Sun-grazer" and would pass perihelion on October 21. Moreover, conditions for observation at and immediately prior to perihelion passage would be the best possible for a Sun-grazing comet—in the hours preceding perihelion passage, the comet would be at minimum distance from Earth and maximum elongation from the Sun for an object of such small heliocentric distance. Even so, it would necessarily be extremely close to the Sun, but in all probability it would become so bright that unique observations of a Sun-grazer passing through perihelion could be made.

Early orbital calculations were not sufficiently accurate to enable astronomers to determine just how close to the Sun the comet would pass, and estimates ranged from around 1.3 million kilometers down to a direct collision!

Rather surprisingly, no prophet of doom (to my knowledge anyway) exploited the collision prediction. (Of course, there were no real grounds for any doomsday talk, but this in itself is no real drawback to those who wish to indulge in morbid pastimes.) In any case, more accurate orbital determination during the first week of October showed that the comet was not destined to plunge into the Sun, but would pass about 480,000 kilometers above the photosphere—certainly well within the corona, but not sufficiently close to be in danger of collision.

Meanwhile observations during early October revealed the comet to be daily increasing in magnitude, and photographs recorded a typical type I gas tail with several bright streamers visible. By midmonth, the head had reached second magnitude or brighter, and some 10 degrees of tail were visible at times with the naked eye, although observations were being made increasingly difficult by moonlight and twilight, and as the date of perihelion approached, the ever-brightening comet was also proving to be ever more elusive. Just eighty hours before perihelion, observers at Woomera could not detect any tail at all in the bright sky, although they estimated the comet itself to be equal to a star of zero magnitude.

As the comet neared the Sun, the size and magnitude of the visible nucleus increased relative to the coma. During late September, the coma was about 3 or 4 minutes of arc in diameter and of magnitude 6 or 7, the nucleus at this time being a mere starlike point of magnitude 9. However, by October 14, the coma had shrunk to only 0.2 minutes of arc, having a magnitude of about 2 or 3, while the nucleus had expanded into a disk some 7 to 10 seconds of arc in diameter and having a magnitude of approximately 4. Near perihelion the comet's head had decreased to only 12 seconds of arc, demonstrating how a comet becomes virtually "all nucleus" at Sun-grazing distances.

By mid-October, floods of rather wild predictions began to be made. Newspapers told the world that a comet brighter than the Moon was on its way, conveniently omitting the fact that this fantastic comet would be so close to the Sun that it would need to be almost as bright as the Moon to be seen even with special telescopes! Unfortunately, even some astronomers became a little carried away and predicted that the tail would "sweep like a giant searchlight" across the sunset skies of the U.S.A. as the comet (located just below the western horizon) swept around the recently set Sun. This would indeed have happened if the tail had remained straight during the million-odd kilometers per hour perihelion sweep—not a very likely prospect as the outflow of material along the tail could not reach accelerations of the order of 1.6 million kilometers per hour, as would have been necessary for this prediction to have reached fruition.

Perihelion day, October 21 U.T., proved a very mixed blessing. In many places early risers saw only cloud and haze. That morning (nine hours before perihelion) I had my first view of the comet, after an inconvenient and annoying spell of cloudy weather. From the summit of a small hill overlooking the industrial city of Newcastle (sometimes affectionally known as "the smoky city"—not the best place for viewing comets) the object was located with 20 × 65 and 8 × 32 binoculars, deep within the morning twilight—in fact, only 1.5 degrees from the center of the Sun. It was extremely brilliant and had a tail about 0.25 degrees long. During observation, the Sun rose, but remained safely hidden behind a thick cloud bank on the horizon. I stopped looking, for safety reasons, once the Sun's rim began to appear.

Later in the day, hundreds of people saw the comet with both the naked eye and through optical aid, especially in the central U.S.A. and Hawaii, where it was described as being a conspicuous naked-eye object very close to the Sun. Across the Pacific, in Queensland, one observer watched it through welder's goggles when within 0.5 degrees of the Sun, and in the U.S.A., David Meisel made naked-eye drawings of the comet as seen near the Sun in the daytime sky. He noted some 2.25 degrees of tail visible with the naked eye, even when the comet's head was about 0.5 degrees from the limb of the Sun (Plate 11).

Gerald de Vaucouleurs, at the McDonald Observatory in Texas, observed the comet some ten hours before perihelion when

Plate 11. *Naked-eye daylight drawings of Comet Ikeya-Seki (1965 VIII). As the comet moved toward perihelion, Dr. David Meisel made these sketches from Flagstaff, Arizona, on October 20, 1965, at 2100 hrs. U.T. (the sketch on the left) and at 2300 hrs. U.T. About 2¼ degrees of tail were visible.*

it was at an elongation of 2 degrees and estimated its brightness to be as high as —10, commenting that this was about one magnitude brighter than had been predicted. Half an hour later, Norbert Roth and Darrell Fernald saw it from the Smithsonian Station at Organ Pass, New Mexico, noting a 1-degree tail as bright as the 25½-day-old Moon visible in the sky at the same time. These estimates were confirmed by independent observations made by Elizabeth Roemer (Flagstaff Observatory) when, four hours prior to perihelion, she reported the comet to be of magnitude —10 or —11 with a markedly curved 2-degree tail.[19]

In Japan, the comet was observed by a number of people in daylight (including one of its codiscoverers, T. Seki), and it was photographed at Tokyo Observatory's Mount Norikura coronagraph station (Plates 12 and 13) during the actual perihelion passage and for some time thereafter. Some of these observations (made with a 122-mm telescope) must have been obtained when

Plates 12 and 13. *Comet Ikeya-Seki (1965 VIII) photographed from the Mount Norikura solar station of the Tokyo Observatory, F. Moriyama and six co-workers obtained these photographs at 0202 hrs. U.T. (Plate 9) and 0327 hrs. U.T. (Plate 10), October 21, 1965, with the aid of the 11.9-cm coronagraph in 4,700–6,000-angstrom light. The Sun is behind the black disk which is fringed with scattered sunlight.*

the comet was only some 10 minutes of arc from the Sun's limb, and yet the tail was still clearly visible. The brightness of the comet continued to increase, according to these coronagraph photographs, and it was described by Mount Norikura astronomers as being "ten times brighter than the full Moon" around perihelion —a description which indicates magnitudes in the range of −15 or −16! With this brilliance emitted from the very small disk of the head, the intensity of light must have been close to that of the Sun's limb itself, and it is perhaps not too surprising that photographs could be obtained under these conditions. Nevertheless, this was a remarkable feat of observational astronomy, and it allowed us unprecedented views of a comet under the most extreme of conditions.

Just half an hour before perihelion, a disruption in the head was noted, but the comet survived perihelion passage at the price of being broken into three fragments, one considerably brighter than the others. Disintegration must have been very rapid, as two of these fragments had disappeared when the comet became generally visible again some hours later.

Following perihelion, there is a gap in the observational record lasting for at least six hours. During this time, the comet passed within a quarter of a minute of arc of the limb of the Sun (being then beyond the Sun) and would have been a very difficult object to detect, but considering the high brightness of the object just prior to perihelion, it is perhaps surprising that *no* observations were obtained during this period.

Evaporation of small particles must have been very rapid at this time. Indeed, when the comet again became visible in the daylight sky, the tail was found to be considerably weaker than it had been during the pre-perihelic section of the orbit. Likewise, calculations of syndynes and synchrones reveal that the spectacular tail, which became so prominent in the weeks following perihelion, was composed of particles released during the post-perihelic section of the orbit.[20] Earlier particles simply could not have survived the inferno through which the comet plunged at perihelion.

Several hours after perihelion, the tail was already growing stronger again and the comet had become a daylight naked-eye object of magnitude −6 or −8. Indeed, it remained a telescopic daylight object for two or three days following perihelion and, apparently, could still be detected as late as October 25, although by this time it was becoming of greater interest as an early morning object with a head estimated to be as bright as magnitude —2 and a tail some 20 degrees in length and 3 degrees wide at the extremity.

It soon became evident that the comet was exhausting, but it was determined to go out in a blaze of glory! Material boiled away from the nucleus could not simply vanish into thin space; it had to go somewhere—"somewhere," of course, being the rapidly growing tail.

By October 28 the new dust tail was a commanding and beautiful sight in the still dark predawn sky. Stretching for some 30 degrees (even 45 degrees according to Mr. Minton, who observed the comet under exceptionally favorable conditions) the real length of the tail was a full 1.3 A.U., that is to say, considerably longer than the distance of Earth from the Sun. The first 20–25 degrees of tail were exceptionally bright.

In late October, a faint anti-tail was also observed (by Minton and others), as well as a faint secondary streamer to the south of the main tail. Both these features were also photographed. The main tail showed a bright central core and, in the center of this, a dark lane visible for most of its length (a feature also noted in Comet 1882 II). The tail was also somewhat unusual in that it did not grow fainter with distance from the head. Rather it became noticeably brighter at a distance of about 14 degrees from the head, remaining uniformly bright for another 7 or 8 degrees and then growing faint fairly rapidly. It is instructive to note that the brightest section of the tail marked the beginning of a series of oblique striations, presumably pseudosynchronous features similar to those observed more recently in West's Comet of 1976. This observation would appear to add support to Sekanina's hypothesis concerning the nature of these features, as we discussed in Chapter 1. Possibly related to these striations, an unusual feature was noted on a photograph taken by Bester at Boyden Observatory on November 2—a very faint glow extending like a spectral finger, at a considerable angle to the tail axis, emerging from the main tail some 4 degrees from the head.

Unfortunately, after the "disappointment" at perihelion, the news media largely dropped the comet, a pity in view of the spectacle which occurred ten days after perihelion passage. At that time the comet could be seen against a still completely dark sky, but it had not yet traveled sufficiently far from the Sun to be too faint for a great show. On the morning of October 31, in an exceptionally clear and dark sky, I was fortunate in having had a "grandstand" view from a small headland away from artificial lights. At 3 A.M. the sight was magnificent beyond description. So dark was the sky in the east, that sea and sky both merged in an invisible horizon, from beyond which the comet head had just emerged. Like a brilliant white arc, the great tail curved upward for nearly 30 degrees, spreading as the distance from the head increased, yet not waning perceptibly in its intensity. Indeed, so intense was the tail that the coma itself was hardly distinguishable (as a separate entity) with the naked eye, even though its brightness was still as high as second magnitude.

Fortunately, I have been able to observe both this comet and Bennett's (see below) under near-perfect conditions when at their finest, and I am certain that Ikeya-Seki was the more spectacular of the two. Yet far fewer people saw it, due no doubt primarily to its brevity of display and the unfortunately inaccurate notice given it by the news media. This is a great pity, as it was surely the finest comet for many years, and it may be a very long time before one as spectacular comes our way again.

As the comet moved farther from the Sun, the signs of disintegration became more pronounced. Soon after perihelion, the central condensation became fuzzy, probably due to the release of a cloud of meteoric dust which obscured the true nucleus. Early in November, the nucleus itself was seen to be multiple, having two main components (the smaller of the two being itself divided into a very close triplet) and a third, much fainter one possibly observed some distance from the other two. Other additional nuclei, perhaps very transitory in nature, were reported by a number of astronomers. These nuclei apparently originated near perihelion but do not seem to have been associated with the fragments observed by Japanese astronomers at the time of perihelion passage itself—such fragments probably had a status similar to the temporary "satellites" of 1882 II, although they did not survive long enough to develop nebulous comae.

During November, the comet's head became larger, more diffuse, and of lower surface brightness, making magnitude estimates very difficult and highly dependent upon atmospheric conditions. For instance, on November 26, Mr. G. Solberg estimated the comet to have fallen to magnitude 7.5 and the tail length to have decreased to 15 degrees. However, on that morning I was able to observe the comet and about 15 degrees of tail (at least) with the naked eye—the head, in fact, appeared to be of greater size (as seen with the naked eye) than it had some ten days earlier. It seemed to me to appear somewhat similar to the globular star cluster 47 Tucanae. On that morning also, Mr. John Davies of the Blue Mountains, west of Sydney, estimated the total brightness to be as high as 3 or 3.5 and traced a tail of at least 20 (possibly even 30) degrees length with his unaided eyes! On the other hand, Mr. Alan Marks, observing the comet at the same time, but from within the metropolitan area of Sydney, could not see it without optical aid. All of this suggests that most of the comet's light came from the extended outer coma, a region of quite high total light but low intensity and requiring good conditions to be seen to advantage. This phenomenon strongly hints at the progressive exhaustion of the comet.

When these observations were being made, the comet was located some 1.02 A.U. from the Earth and 1.2 A.U. from the Sun. In early December it had receded to 1.03 A.U. from Earth and 1.4 A.U. from the Sun but was still a naked-eye object. My last naked-eye observation came toward the end of the first week of December, at which time the comet was almost overhead at dawn and showed a tail visible for at least 3 or 4 degrees with the naked eye, in a dark sky. The head, however, could not be distinguished without the aid of field glasses, in which it appeared as a very diffuse globule of mist, uncondensed and featureless. So low was the surface brightness of the tail that much of its length disappeared under the low magnification of small field glasses. By mid-December, the comet appeared as a rapidly fading patch of light, visible only in the wide field of a pair of powerful binoculars, and then only as an area of light sky, almost half as large as the full Moon. Far fainter than predicted in January 1966, it was last observed with certainty by Tammann of Palomar Observatory on January 14 (when both nuclei were still visible), although there was a further possible observation recorded on Baker-Nunn plates at the Smithsonian stations as late as February 12.

Bennett's Comet (1970 II)

On the night of December 28, 1969, John Bennett of South Africa discovered an 8.5-magnitude uncondensed comet near the Magellanic Clouds (which, I may also mention with mixed emotions, was discovered independently by myself less than four days later). Little did anyone guess that this inconspicuous object moving at a snail's pace through the south circumpolar skies would, within three months, evolve into one of the major comets of the century and be compared by some observers to such historic objects as P/Halley of 1910 and Donati's Comet of 1858!

During the coming weeks, the blob of nebulosity moved north, becoming increasingly condensed and sprouting a rudimentary tail. Initial orbital determinations revealed that the comet would reach perihelion on March 20, 1970, at which time it would be little more than 0.5 A.U. from the Sun. Furthermore, its orbit was such that at perihelion it would be well placed for observation, being located in the equatorial regions of the heavens at considerable altitude in the predawn sky. Hopes were therefore high, as magnitude predictions indicated a brightness of at least second magnitude and, assuming that the comet behaved normally, a long bright tail could confidently be expected.

The comet's brightness increased even more rapidly than anticipated, and by the end of the first week of February it could be seen with the naked eye if the observer knew exactly where to look. By February 9 the head appeared very condensed and a fairly faint tail swept across the 3-degree field of my 20 × 65 binoculars when averted vision (looking slightly to the side of the comet so that its light falls on the more sensitive part of the eye) was employed.

As February passed into March, the comet became invisible in the evening sky but increasingly easy to observe in the southeast before dawn. A bright curving dust tail several degrees long had become visible with the naked eye by March 6, the initial faint gas tail emerging from the convex edge of the brighter one about 1 degree from the head. These two tails were visible almost to the time of perihelion, when the main tail eclipsed the secondary one for a few days, the latter emerging again from the convex side of the dust tail (which, for southern observers, had now become the edge nearest the horizon) by April.

In a very dark rain-washed sky on the morning of March 21, the comet was a beautiful sight in the eastern sky. The broad, pale yellow dust tail was about 13 or 14 degrees in length (as estimated with the naked eye), with a sharp convex edge facing north and an indefinite, diffuse, concave edge blending into a pale yellow glow to the south, giving the appearance of a yellowish fountain curving back away from the comet's direction of motion and dissolving into a fine mist of luminous yellow spray. In the days following, the (now full) Moon hardly diminished the comet's splendor as the brilliant tail continued to lengthen.

Through a small telescope, the comet's head displayed a markedly parabolic outline, having a brilliant yellow starlike nucleus almost at the focus of the parabola. Larger telescopes and powerful eyepieces revealed a wealth of detail in the near-nuclear region of the coma, including a system of jets and envelopes reminiscent of those noted in the head of Donati's Comet of 1858 and Halley's Comet of 1910, in addition to the narrow, dark "shadow of the nucleus" also prominent in Donati's Comet.

An especially interesting feature of the near-nuclear region of this comet was the display of "orange pinwheels" (as they were graphically described at the time) giving the impression that the nucleus was experiencing rather rapid rotation.

Rapid variations and a high level of activity were also evident in the (at times highly contorted) gas tail, apparently due to rapid and rather sudden variations in the production of ionized carbon monoxide, its major constituent. In contrast, the main tail showed little variation in intensity but increased steadily in length, reaching some 25 degrees by April 11, according to Dr. R. L. Waterfield at Woolston.

The comet remained a naked-eye object from early February until early May, reaching a maximum magnitude of about zero at perihelion—somewhat higher than initially expected. Thereafter it became a telescopic object for the remainder of the year, being last observed on February 27, 1971, some fourteen months after discovery, when at a distance of over 5 A.U. from the Sun.

Kohoutek's Comet (1973 XII)—The Great Comet That "Failed"

On the evening of March 7, 1973, Dr. Lubos Kohoutek of Hamburg Observatory photographically discovered his second comet in less than a week—a tiny sixteenth-magnitude blob having neither condensation nor tail. Later, he found that its image had actually been previously recorded on a plate taken on January 28.

Subsequent orbital calculations revealed that the new object was still almost 5 A.U. from the Earth and Sun, and that it was not due to arrive at perihelion until December 28, at which time it would pass the Sun at a distance of little more than 0.1 A.U. Naturally (other things being equal) the comet's magnitude was expected to increase markedly as it approached perihelion; but the exact value of maximum brightness could not be estimated accurately, and some estimates put forward in the early days following its discovery suggested that the comet could be brighter at perihelion than one would reasonably expect.

The comet at discovery was not exceptionally bright for an object at such a distance (several objects of similar, sometimes brighter magnitudes have been observed at these distances in recent years), but early predictions assumed a high value for the parameter n—i.e., they assumed that the comet was increasing in brightness at a higher-than-normal rate. Initially, indeed, this seemed justified, and the comet's light curve could be represented by a value of n near 8 and a corresponding absolute magnitude in excess of —2, which would have implied impossibly high brilliance at perihelion. Nevertheless, such high values of n are not representative of large new comets and are more likely to be found among small, gassy, and diffuse objects of short period.

Unfortunately, it was these rather unrepresentative predictions which reached the general public via the news media. The most popular was a prediction based upon the (rather unwarranted) assumption that the comet would continue to increase in brightness according to $n = 6$ and $mo = 2.5$, giving a magnitude of —10 at perihelion, a magnitude probably only reached by Sun-grazing comets, and a value still as high as —6.2 on January 3, 1974, some six days after perihelion when the comet would have been visible against a fairly dark evening sky. Such a sight would have, indeed, been unbelievable—which is probably a very good reason why many astronomers did *not* believe it!

Certainly there was no guarantee that the comet would continue to increase in brightness at this rate (if, indeed, it could even be shown that the comet *was* increasing at such a rate, with brightness estimates taken off photographic plates and covering only a very small arc of the comet's orbit being all that was available), and even if it *did,* there was every possibility that it would depart from this at perihelion, when rather extreme conditions near the Sun would be encountered.

Maybe with such thoughts in mind, various astronomers proposed more moderate predictions. For example, S. W. Milbourn, then Director of the Comet Section of the British Astronomical Association, predicted a perihelion magnitude of —2.5—a surprisingly accurate (even, possibly, a little conservative) prognostication, but one to which very few people listened at the time.

Such "conservative" predictions were soon forgotten in the wave of hysteria (that is not too strong a term) which followed, especially in the U.S.A. where certain commercial interests were served well by the wave of "comet mania," with sales of everything from telescopes to Kohoutek T-shirts commemorating the "Comet of the Century" (an item of clothing still for sale in 1974, but by then said to be commemorating the "Scientist's Comet").

Unfortunately the popular press was not the only sensationalistic element. Even a noted and sober scientific journal published an article in which it was claimed that the tail may "span the entire vault of the heavens." A little trigonometric calculation would, however, have dispelled this hope as (when phase angle and general perspective were taken into account) it would have been realized that the tail would need to be impossibly long (in real terms) before it could span anything like "the entire vault of the heavens."

While the sociological phenomenon built up on Earth, the astronomical one which was causing the whole furor drifted slowly into the oblivion of the sunset glare, a tiny spot visible only on photographs taken with giant telescopes until it was lost completely in the twilight during May.

After vanishing from the evening sky, the comet did not reappear until late September, when T. Seki recovered it as a diffuse object of magnitude 10.5 or 11, low in the predawn sky. This was slightly fainter than had been predicted, but the discrepancy alone was not serious. Rather more ominous was the discovery (from the determination of an improved orbit based upon a wider arc of positional measurements) that the orbit was that of a "new" comet in the Oort-Schmidt sense, that is to say one which is making its first trip to the region of the Sun after deflection from the circumsolar cometary cloud. This discovery was ominous because such objects usually become active at considerable heliocentric distances and brighten at a slower rate than those objects which have passed the Sun on previous occasions.

Be this as it may, a tremendous scientific reception had been planned for the new visitor, and this was ready to get underway. Not only were there ground-based observations (some of which were of a new and accurate kind—e.g., stellar occultation experiments and radio astronomical observations), but also aerial and extraterrestrial observations were planned on a scale never before attempted.

Aerial observations of comets have a fairly long, if very broken, history. Balloon observations were suggested for Schaeberle's Comet of 1881, and in the following year an astronomer in Paris drifted in a balloon to make daytime observations of 1882 II from above a layer of clouds. Balloon observations were also made by Russian astronomers in 1926 (of Ensor's Comet), but intensive aerial comet observation had to wait until 1965, when jet aircraft, converted into flying observatories, were used for the study of Comet 1965 VIII.

The first faltering steps in "rocket cometology" were also made in 1965, with the planned launch of three Aerobee rockets, two from Wallops Island (Virginia) and one from White Sands (New Mexico). Unfortunately, only the White Sands flight succeeded, and this apparently failed to produce any revolutionary results, although a spectrogram was obtained. Real success had to await the arrival of the two bright comets of 1969–70 (Tago-Sato-Kosaka and Bennett) when satellite observations revealed the existence of the huge, and thitherto unknown, coma of hydrogen.

Such observations were to be extended, in the Kohoutek program, to include scrutiny from the Skylab space station and the Mariner probe to the inner planets.

In addition to the number and accurate nature of the planned experiments, the advance notice given by the early discovery of the comet enabled the coordination of experiments and observations on a scale never before attempted. More than anything else, this advance planning made this comet such an important object for astronomers interested in cometary research.

Unfortunately, the memories of casual amateurs and the general public are not as enthusiastic. By December, it was becoming obvious that the comet would be much less brilliant than had generally been anticipated. Not only had its magnitude increased much more slowly than had been predicted, but the *rate* at which it was increasing was actually falling. That is to say, it was increasing in brightness, but at an ever decreasing rate (or expressed in terms of the brightness formula, the value of n was decreasing as the comet approached perihelion).

Nevertheless, keen-sighted observers could spot the comet with the naked eye from about November 20 if they had a sufficiently clear sky and knew exactly where to look, and by early December a hazy tail approximately 1 degree long made the comet an impressive sight in binoculars.

Predictions concerning the tail length were made for this comet, and they proved to be rather more accurate than the brightness prognostications—except, that is, for a period around December 17, when tail growth became so rapid that it actually attained a length of some five times in *excess* of the predicted value, extending for a full 15 degrees in the morning twilight, as seen with the aid of wide-angle field glasses.

By then, however, the comet was rapidly becoming a difficult visual object in the encroaching dawn. At no time did it become a conspicuous early morning object as had been hopefully predicted.

Apparently, the comet experienced an outburst in brightness near perihelion, as observations from both the Orbiting Solar Observatory (on December 26 and 27) and the Skylab manned space station indicated the object to be considerably brighter than suggested by its behavior in mid-December. The maximum magnitude was at least −1.5 and possibly as high as −3, ironically making the predictions that the comet would become far brighter than Halley's perfectly correct—but the Sun was so close that the comet could not then be seen from the ground!

This high level of activity was only brief, however, and was compensated for by a fall in intrinsic brightness to a level about one magnitude *below* that of early December.

Nevertheless, the comet was widely observed as it emerged from the solar rays and into the western evening sky. Thanks to an abnormally cold, clear, air mass over the southern U.S.A., the comet was there observed with the naked eye throughout January as a delicate and beautiful wraithlike object. January 15 saw the comet displaying a broad curving dust tail about 10 degrees long, together with a straight narrow gas tail extending some 25 degrees, according to photographs obtained at South Baldy Observatory in New Mexico. On that evening, astronomers at the observatory could see some 16 degrees of tail with the naked eye, the comet's magnitude being about 5. These observed tail lengths are very close to the predicted maximum dimensions, for mid-January, as calculated several months earlier by W. Liller and independently by Z. Sekanina.

Though hardly a "great comet" in any sense of the term, Kohoutek was certainly a striking naked-eye object and one of the most scientifically significant ever seen. A wealth of new material was discovered during observations of this object—for instance, the discovery of complex molecules (HCN, CH_3CN—hydrogen cyanide and methyl cyanide) by means of observations at radio wavelengths and confirmation of the presence of H_2O^+ (ionized water) in the comet's tail, the first direct indication of the presence of water in comets. Study of the anti-tail which developed after perihelion also provided a good deal of information regarding the motion of large dust particles in comets and spearheaded the revival of interest in these features in recent years. In fact, one only needs to read any journal article or paper on comets written since 1974 to see the extent to which modern cometary astronomy has been affected by the comet many people still class as a failure!

During early 1974, the comet faded rapidly, moving into the evening twilight and finally becoming invisible after April 26.

Hopes that it might be reobserved as a very faint object later in the year were not, apparently, fulfilled.

Nevertheless, as it faded it was joined in the evening sky by another object which, at least in my opinion, provided an even more spectacular sight for observers with good binoculars. I refer, of course, to Bradfield's Comet (1974 III), an object which reached about fourth magnitude to the naked eye and displayed a bright tail several degrees in length. Though not attracting the publicity of Kohoutek, this comet was an important object as it became the first in which H_2O (as distinct from ionized water—H_2O^+) was observed, thereby adding still further support for the icy nature of these objects.

The comet was also interesting in that it passed very close to the crescent Moon (in angular terms only, of course), making a very impressive sight on March 26. Though it may have passed unseen, an occultation of part of the tail actually occurred that evening—visible from Siberia.

Moving north, the comet passed only 0.6 degrees from the north celestial pole on May 14–15, by which time it had faded to magnitude 9 but still displayed a tail of some 20 minutes of arc in length.

West's Comet (1976 VI)—The Unexpected Great Comet

This object was initially found in November 1975 on plates taken at the European Southern Observatory during the previous September as part of the "quick blue" survey of southern stars.* Subsequently, the comet was found on plates taken in August, and a sufficient number of positions could be derived from the trailed images to enable orbital computations to be made.

Now faint comets are not infrequently discovered on survey plates, and those for which an orbit can be calculated are almost invariably found to be either very distant objects or else short-period comets of very low intrinsic magnitude.

A few are observed once or twice only and slip away without

* The "quick blue" survey is the term used for the initial E.S.O. blue-light photographic survey of the southern sky.

an orbit's ever being calculated. Therefore, we can imagine the surprise when this new object of magnitude 16 or 17 was found to be moving in an orbit having a perihelion distance of just under 0.2 A.U., which it would pass on February 25, 1976. It was almost Kohoutek all over again!

Clearly, the comet would experience a considerable increase in brightness as it neared the Sun, but it did not appear especially bright on the survey plates and nothing very spectacular was expected, although a predicted magnitude of about 5 in mid-March and a favorable position in the morning sky for northern hemisphere astronomers made the object appear an interesting prospect.

An ephemeris for the new object was published, and some more observations were obtained in confirmation of the general accuracy of the orbit, but the comet remained faint and did not attract a wide range of attention. It was not until late December and early January that the first visual observations were obtained, and these, a little surprisingly, made the comet nearly two magnitudes brighter than predicted. Nevertheless, even this did not raise hopes too high for a spectacular display in March, as the tail was faint and relatively unimpressive even in late January, and astronomers (very wisely) cautioned against assuming that this higher intrinsic brightness would necessarily continue through perihelion passage.

By mid-February, it was beginning to look as though this caution would prove correct as the magnitude estimates for this period were very close to the original prediction, apparently indicating that the high initial values (in January) were not going to be representative of the apparition. On the evening of February 14, the comet seemed to me to be about third magnitude and had already developed the strong yellow color which was to become a notable feature of this object. The coma had shrunk to an almost starlike point, and a faint fan-shaped tail had become visible, although this could not be seen clearly in the bright evening twilight.

Nevertheless, just as it appeared that the comet might not become exceptionally brilliant, two amateur astronomers (Richard A. Keen and Alan Hale) located the comet just after sunset on February 21–23 as an object of magnitude −1—once again, two magnitudes brighter than predicted!

This time, the new high brightness was maintained. Only seventeen hours after the comet passed perihelion, Mr. John Bortle picked it up with the naked eye ten minutes *before* sunset, at which time he judged the magnitude as −3. In 10 × 50 binoculars, he estimated the tail of the daylight comet as 0.6 degrees, and similar daytime estimates were made by a number of other astronomers on this day. Even as late as February 27, Mr. Bortle could still detect the comet in daylight with the aid of 15 × 80 binoculars and estimated the magnitude to be −2.4.

In early March, the comet emerged into the morning sky, having a magnitude of about −1 and a strong yellow color. It rapidly became well placed for early morning observers in the northern hemisphere and showed two tails—a bright and spectacular dust tail (which displayed interesting pseudosynchronous features) and a (initially) fainter, bluer gas tail. By March 7, the main tail was as long as 30 or even 40 degrees, according to Daniel Green at North Carolina. On this morning, John Bortle counted no fewer than five distinct tails, up the 25 degrees in length, and remarked that the general appearance of the comet was quite reminiscent of descriptions and drawings of the Great Comet of 1744.

On March 8, Mr. Bortle saw the nucleus as clearly double, and by March 12 four distinct condensations were visible. This division of the nucleus was carefully watched by professional astronomers in an attempt to learn more about the structure of these objects, and the results certainly appear interesting and tend to confirm the orthodox icy-conglomerate model.

Soon after the schism was noted, the subnuclei appeared somewhat similar to the famous Trapezium multiple star system in Orion, but this formation soon disintegrated as two of the fragments (one in particular) underwent increasingly powerful nongravitational accelerations. The nucleus which displayed the most powerful acceleration separated from the group rapidly during March, but at the end of the month it experienced a sudden drop in brightness and seemed to dissipate completely after March 28. The other nucleus which showed a certain degree of acceleration also displayed rather large variations in magnitude, actually becoming brighter than the main nucleus on March 29 before fading sharply to a value clearly below the other two remaining nuclei in April.

This behavior is probably instructive, as it strongly implies that the nuclei having the strongest acceleration were relatively small and icy objects which quickly disintegrated. Such small fragments would (according to the icy-conglomerate theory of Whipple) be expected to be the most severely affected by the thrust of escaping gases. They would also be prone to rather large fluctuations in brightness as the ices composing most of their mass crumbled and evaporated.

The fact that West's Comet experienced such a major schism at a distance of at least 0.2 A.U. from the Sun, while other comets such as Seki-Lines (1962 III) pass as close as 0.03 A.U. without any indication of disintegration indicates the range of properties possessed by these objects. Presumably, West's was an unusually large and fragile comet, easily upset by its close approach to the Sun. Presumably also, this very fragility was closely linked with its spectacular display—i.e., with the high productivity of fine dust constituting its large tail. A comet of greater density and less fragility—as, we must suppose, Kohoutek's (1973 XII) to have been—would almost certainly not have provided the spectacle displayed by West.

As the comet moved away from the Sun, the dust tail faded fairly quickly and had already become fairly faint by March 13, whereas the gas tail remained strong and easily observed.

The total magnitude of the comet faded more slowly than had been anticipated, and it could still be detected with the naked eye during the first half of April, fading to about magnitude 6.6 by that month's end (according to Mr. Bortle's estimate with 10 × 50 binoculars). Even at the end of July, the comet was still accessible to small telescopes as a very diffuse object of approximately ninth magnitude.

In the years to come, West's Comet will probably be the most remembered of the large objects of the latter half of this century, as it not only attained high brilliance and developed a spectacular tail (or tails!) but was very well placed for observers in northern latitudes—and it is in northern latitudes where most astronomers still live, irrespective of the fact of large new southern observatories. West's Comet was larger and brighter than Bennett's and was better placed (for most people, though not for observers at my own latitudes) than Ikeya-Seki (1965 VIII), and the photographic coverage by both amateur and professional astronomers was unprecedented in the number and quality of portraits obtained.

Of course, the future may *not* remember West's as the prime comet of the latter half of the twentieth century. Perhaps there will appear, before the century is over, another great comet even more spectacular and even more favorably placed for the world's observers. This is something which no one knows—yet even if there are no more great comets this century, there will be one famous object returning in the mid-1980s which, though hardly expected to become a great comet, will certainly arouse much interest and comment. I refer of course to the most famous of all comets, the one which (for many people) is *the* comet—the periodic comet of Halley.

REFERENCES

1. G. F. Chambers. *The Story of the Comets.* Oxford: The Clarendon Press, 1909, p. 126. (Grant's list originally put forward in a lecture to the Royal Institution in 1870.)

2. S. K. Vsekhsvyatsky. *Physical Characteristics of Comets.* Published for NASA and the National Science Foundation, Washington, D.C.: Israel Program for Scientific Translations, 1964, pp. 47–81.

3. R. A. Lyttleton. *The Comets and Their Origin.* London: Cambridge University Press, 1953, pp. 52–53.

4. F. G. Watson. *Between the Planets.* The Natural History Library, Garden City, N.Y.: Doubleday-Anchor, 1962, p. 73.

5. S. K. Vsekhsvyatsky. *Physical Characteristics of Comets,* p. 124.

6. Ibid., p. 167.

7. Ibid., pp. 130–31.

8. Joseph Ashbrook. "The Great September Comet of 1882." *Sky and Telescope.* Vol. 22 No. 6, Dec. 1961, p. 331.

9. S. K. Vsekhsvyatsky. *Physical Characteristics of Comets,* p. 265.

10. R. S. Richardson. *Getting Acquainted with Comets*. New York: McGraw-Hill, 1967, p. 183.

11. G. F. Chambers. *The Story of the Comets,* p. 153.

12. B. G. Marsden. "The Sungrazing Comet Group." *The Astronomical Journal*. Vol. 72 No. 9, Nov. 1968, pp. 1170–83.

13. W. T. Lynn. *Observatory*. Vol. 11, 1888, p. 375.

14. C. M. Botley. *Sky and Telescope*. Vol. 33 No. 2, Feb. 1967, p. 84.

15. W. R. Brooks. *Popular Astronomy*. Vol. 23, 1915, p. 449.

16. B. G. Marsden. "The Sungrazing Comet Group."

17. Ibid.

18. Ibid.

19. B. G. Marsden. "The Great Comet of 1965." *Sky and Telescope*. Vol. 30 No. 6, Dec. 1965, pp. 332–37.

20. B. J. Jambor. "The Split of Comet Seki-Lines." *Astrophysical Journal*. No. 185, 1973, pp. 727–34.

4

Halley's Comet—The Most Famous of Them All

The year 1682 saw the arrival of what was to become the most important, if not the most spectacular, comet in the history of the dynamical study of these objects, as this was to be the first periodic comet definitely recognized as such. It is a little difficult for us today, with periodic comets visible constantly in our telescopes, to appreciate fully just how revolutionary the concept must have been at the time. Remember, this was the time when comets were thought to be rare and mysterious phenomena, when astrological interpretations and the belief in celestial portents continued to hold sway over people's appreciation of these objects, and when little similarity could have been imagined between

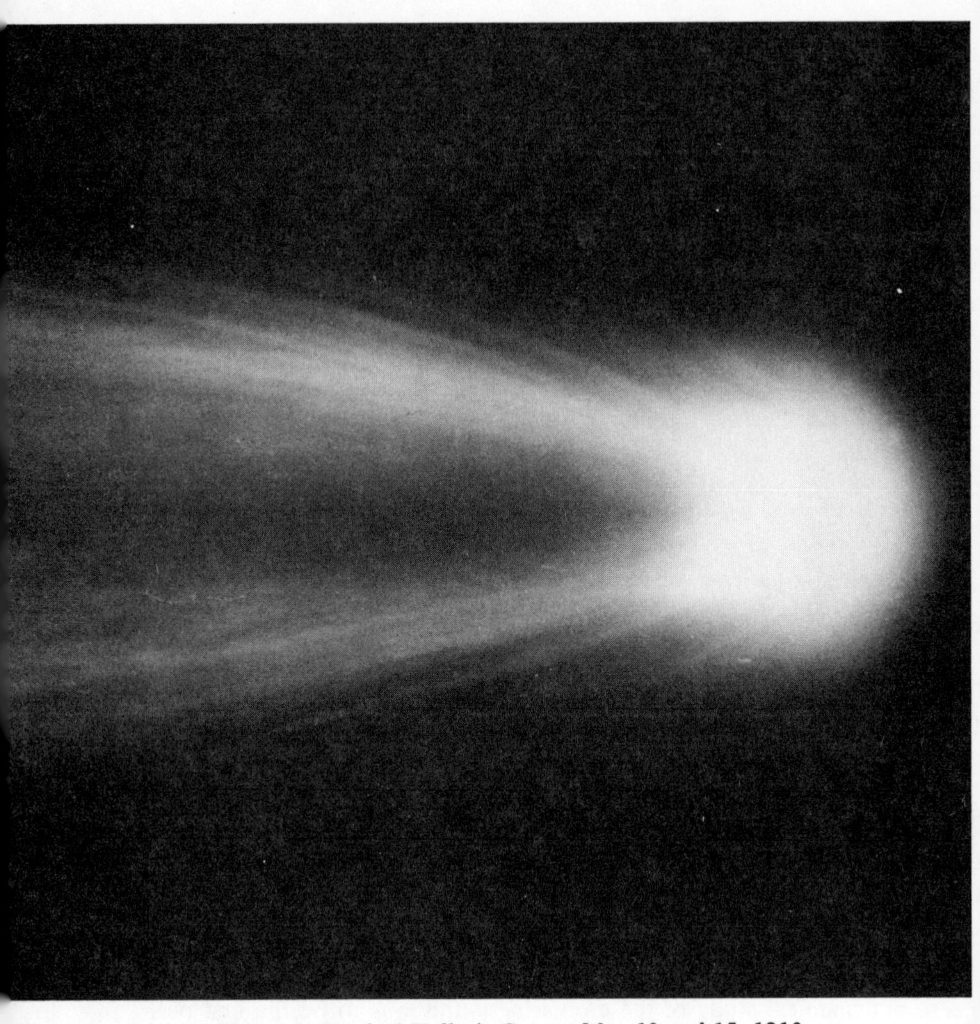

Plate 14. *Head of Halley's Comet, May 12 and 15, 1910.*

the wild and unpredictable comets and the sedate and regular planets. We must remember also that even the latter were only recently being brought within the domain of natural science, and this in itself was a revolution in thought.

The story of Edmond Halley's observations of this comet and his computation of its orbit are well known. Essentially, his method was the same as that of astronomers today—calculation of orbital "elements" and comparison of these with the elements of other comet orbits in an attempt to find any similarity. Of course, today the positional measurements are often to a high degree of accuracy, and electronic computers assist in computing orbital elements in a fraction of the time consumed by the manual methods of former times. Also, our powerful telescopes are able to follow comets over a much longer orbital arc, and this enables elliptical orbits and periods of revolution (where applicable) to be calculated without the necessity of relying on a previous apparition of the comet (although for poorly observed objects, or for those of longer periods, we are not in a position greatly superior to Halley's).

Halley, without any of the modern luxuries or even any precedent to the effect that what he was saying was even possible, nevertheless affirmed that the comet of 1682 was the same object that Kepler and Longomontanus had observed in 1607 and was also identical with the one observed by Apian in 1531. Furthermore, as one final proof of the identity of these objects, records revealed that a bright comet also appeared in the year 1456. Clearly, these apparitions could be explained by the periodic reappearances of a single object revolving around the Sun in an elongated elliptical orbit and having a period of 75–76 years. Thus, Halley predicted, the comet should again reappear in the year 1758, a prediction which he could not hope to see fulfilled, but which he hoped would be remembered by the astronomers of the day.

Remembered it certainly was; however, as the year 1758 drew toward its close and no comet appeared, several astronomers began to express skepticism about Halley's prediction and began to wonder if comets really are the kind of objects for which predictions may be made. Nevertheless, not everyone was prepared to give up hope. The mathematician Clairaut applied what we today would call the theory of planetary perturbations to the simple orbit and concluded that the gravitational effects of Jupiter and Saturn would delay the comet somewhat beyond the 1758 date, derived from the comet's simple orbit. The comet would not actually reach perihelion, he believed, until mid-April 1759.

Subsequently, both Halley's original and Clairaut's more accurate prediction were effectively vindicated. On Christmas night 1758 a German farmer and amateur astronomer named Johann Palitzsch assured his place in astronomical history by being the first person to recover a periodic comet on the basis of a predicted return. Initially very inconspicuous, the comet became a striking object after passing perihelion on March 13 (U.T.), and by May 5 it had developed a tail some 45 degrees in length.

Since this historic return, the comet has again been visible in 1835 (when it reached approximately first magnitude and developed a tail 25 degrees long) and in 1910, at which return it became an especially spectacular sight from the southern hemisphere. It is expected back again in 1986, but more on this later.

HALLEY'S COMET THROUGH HISTORY

Halley himself was not in possession of accurate records dating back more than a few centuries before his time. Consequently, he was unable to trace the comet backward through history before the return of 1456. However, in more recent years, Oriental records have become available to Western man, and appearances of the comet have now been extended back many centuries into ancient times, though just how far back the records of the comet actually extend is a matter for some debate.

According to M. Kamienski,[1] the comet can be found in records extending back to the third millennium B.C. Indeed, the first entry in the *General Catalogue of Baldet* (modified version of 1960) lists a comet about the year 2315 B.C. and tentatively identifies this with P/Halley. Similarly, a comet allegedly seen about the year 2241 B.C. is identified with P/Halley. Kamienski argues that a fast-moving comet allegedly seen in either 2191 or 2024 B.C. or even later, within the constellation of Capricornus, probably refers to a rather close approach of the comet in the year 2007 B.C. (approximately). Furthermore, he argues that the

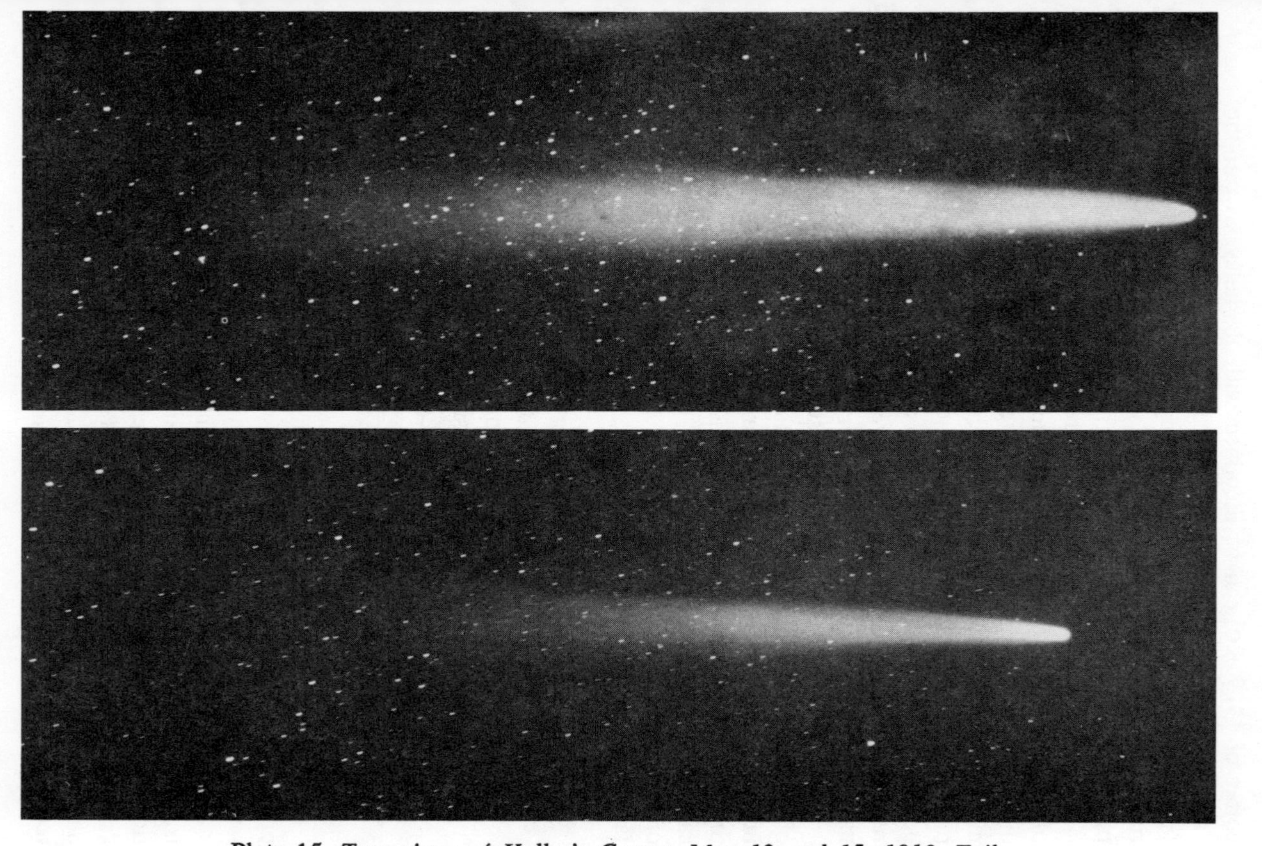

Plate 15. *Two views of Halley's Comet, May 12 and 15, 1910. Tail 30 and 40 degrees long. Photographed from Honolulu with 10-inch-focus Tessar lens.*

comet of 1929 B.C. can be identified with P/Halley, as can the "manifestation of Venus" in the year 1770 B.C. This latter record (preserved by St. Augustine) has long been interpreted as referring to a comet and was (as we have already seen) a favorite suspect for a previous return of the Sun-grazing comet of 1843.

Nevertheless, these early records are clouded with a good deal of uncertainty and are now generally discredited. It is very difficult to derive the correct dates even to within an accuracy of many centuries, and the issue is furthermore confused by the mixing of record and legend. Many of these early "comets" are now believed to have been legendary, and the earliest comet generally recognized as such is usually considered to have been that of 612 B.C. (*not* an appearance of P/Halley).[2]

Nevertheless, no discussion of possible very early appearances of the comet would be complete without mention of the interesting conjecture that Halley's Comet may have been mentioned in the Old Testament. This conjecture is derived from certain passages in the first book of Chronicles, telling of the transgressions of King David concerning the taking of an unlawful census of the Israelites—in particular, the following:

> And God was displeased with this thing; therefore He smote Israel . . . and God sent an angel unto Jerusalem to destroy it . . . And the angel of the Lord was standing by the threshing floor of Ornan the Jebusite. And David lifted up his eyes, and saw the angel of the Lord standing between the earth and heaven, *having a drawn sword in his hand stretched out over Jerusalem.*[3]

The passage then goes on to relate David's repentance and his procuring of the land by the threshing floor of Ornan for the construction of the temple of the Lord.

The part of the above passage which has interested astronomers is, of course, the section I have emphasized in the above quotation. The presence of a comet is suggested by the use of the word "sword"—especially as this term was also used at a much later date to describe an appearance of P/Halley (66 A.D.). For the tentative identification with P/Halley, we have mainly to consult the work of Dr. Gunnar Norlung[4] who, in turn, bases his research upon the orbital derivations of Kamienski. These, according to Norlung's calculations, indicate a return of the comet about the year 1005 B.C., which is probably rather close to the time of the biblical incident (although some Orientalists might well dispute this, placing the incident at a somewhat later date).

Furthermore, Dr. Norlung points out that as the Chronicles account goes on to say that Ornan himself saw the angel *while threshing wheat,* the incident must have taken place in broad daylight. The comet, if that is what it was, must have been a daylight comet and a daylight comet of considerable size and brilliance if it was thought to be a "sword . . . stretched out over Jerusalem!"

The place where the angel was seen is today known as the Dome of Rock (the Mosque of Omar), and if the sword was "over" Jerusalem (in the direction of Jerusalem?), David and Ornan were presumably facing west at the time. Nevertheless, the Sun is nowhere mentioned, which is perhaps surprising if a brilliant daylight comet near the Sun is being implied. But then, could a comet farther from the Sun be as bright as this? It is one thing to see a comet at nighttime as a sword hanging in the sky, but quite a different thing to see a comet in full daylight in this guise.

Indeed, if P/Halley is being described, it is difficult to see how it could have been this brilliant. It has a considerably larger perihelion distance than the brightest of the daytime comets, and unlike them, its light is not from glowing metallic elements evaporated from dust particles under the fierce heat of the Sun. An extremely (perhaps impossibly) high intrinsic magnitude would be implied, and this in turn would imply a large-scale fading since the time of David—a question to which we shall return.

Also, the sword seems to have been of secondary importance. The main "vision" concerned the angel—described in clearly anthropomorphic terms—and it is difficult to see how a comet could be mistaken for a *man.*

In general, then, I think that we must treat Dr. Norlung's interpretation with a good deal of skepticism—a skepticism which must be exercised when dealing with all but the clearest ancient accounts, as it is very easy—when searching ancient records for the existence of some phenomenon—to actually "find" that phenomenon recorded, whether it actually *is* recorded or not.

Still in ancient times, we find comets suspected of having been appearances of Halley in 1175 and 626 B.C. (the second mainly according to Kamienski), and also in 466 B.C., the latter being mentioned in a number of books. Another possible suspect is the comet of 393 B.C., but the comet of 239 B.C. (also widely believed to have been P/Halley) is now considered to have probably been an independent object. Similarly, the supposed identification of the comet of 162 B.C. with Halley is now rejected, and the first object which Marsden identifies with the comet is that of 86 B.C.[5] From then until 1910, the comet has been identified at every return and has had a period ranging from 79.4 years (between the returns of 451 and 530 A.D.) to just under 75 years for the period 1835–1910. In general, the period has been about 76 years.

The various returns of the comet throughout the ages have often been linked with historic events, for which it was, at the time, considered portentous. Thus it was believed to have foretold the death of Agrippa in 11 B.C., the destruction of Jerusalem at its 66 A.D. return, the death of Macrinus in 218, and the defeat of Attila the Hun in 451. The return of 684 was represented in the Nuremberg Chronicle, and the appearance in 837 was probably one of the most spectacular of all. On April 9, it passed 0.039 A.U. of the Earth and moved 60 degrees west in only twenty-four hours! Four days later, Chinese records indicate a tail length of 80 or 90 degrees.[6]

Nevertheless, it is the return of 1066 which is surely the most "memorable" (if that word can be used in such a context), as the comet was then seen as an omen predicting the invasion of England by William the Norman (a good or bad omen depending upon one's allegiance) and was immortalized in the Bayeux Tapestry. According to Chinese records, the comet must have been very brilliant at this return, although it may not have been as bright as the chronicles initially suggest.

Another bright return was that of 1222, during which the comet was regarded as an omen of the death of Philip Augustus of France. According to Oriental records, the comet was visible in the daytime on September 9, presumably under rather exceptional conditions, although a magnitude of at least zero or −1 is implied by this observation.*

The return of 1301 is also interesting historically, as it appears that the comet was represented in one of Giotto's famous Arena Chapel frescoes in Padua.[7] Giotto was noted for the naturalistic style of his work, a style which contrasted with the more normal and largely stylized paintings of his day; and the comet representing the Star of Bethlehem in his Nativity scene reveals details which he must actually have observed in a real comet a few years before creating this work. As the only bright comet during the required time period was the spectacular return of P/Halley in 1301, we can safely regard his painting as a fairly faithful record of its appearance at that apparition.

In 1378 the comet was rather less impressive, but it was described as "terrible, of extraordinary magnitude" at the following appearance in 1456. In 1531 it was observed by Peter Apian, in 1607 by Kepler, and (as we have already noted) by Halley in 1682.

THE COMET'S MOST RECENT RETURN

During its 1910 return, the comet proved to be somewhat disappointing to northern observers, although it provided a truly magnificent spectacle south of the equator. It is difficult to find (at least in my experience) many people who saw both P/Halley and Comet 1910 I, but from the little information I have gleaned, it seems that northern observers considered 1910 I to have been more spectacular, whereas southern observers favored Halley because of the greater length of its tail. There is no question but that 1910 I was by far the *brighter* of the two.

* This estimate is based upon records of the brighter stars having been seen in daylight under exceptional circumstances. If the comet's daylight visibility was *not* primarily due to exceptional atmospheric conditions, its magnitude may have been as high as −3 or −5. Either way (assuming the September 9 record to be accurate), the comet must have experienced a considerable outburst in brightness at that time.

P/Halley was recovered by Wolf on September 11, 1909, as an almost stellar object of magnitude 16. Subsequently, a very faint image was found on a photographic plate exposed as early as August 24, which was not recognized at the time. As the comet moved toward the Sun, the initial signs of a rudimentary tail began to appear about the middle of November when the comet was some 2.7 A.U. from the Sun and shining at about magnitude 12. From November 22 to December 8, there was a series of remarkable brightness fluctuations during which the comet's brightness would temporarily increase from about magnitude 12 to near magnitude 9. At this time, the object was some 2.5 A.U. from the Sun.

The first naked-eye observation (again by Wolf) was on February 11, 1910, and during that month considerable activity was noted within the inner coma. At certain times, the nucleus appeared to have broken into several fragments and the comet was thought to have been breaking up; however, this is now believed to have been unlikely, and what appeared to be secondary condensations were more likely jetlike knots of luminous material erupting from the nucleus, probably related to the jets observed in March when the central condensation had reached a magnitude of about 8. The tail had grown somewhat by this time, and a typical ray structure was becoming evident.

These effects intensified as the comet neared perihelion (April 20) and reached a high level of activity by the end of that month. The tail attained to a length of 6 degrees by May 4, and exceptionally vigorous activity was later noted in both the head and tail.

On May 18, the comet's head passed in transit across the solar disk, although nothing could be observed. Nevertheless, just prior to this conjunction, the tail is said to have extended some 120 degrees across the sky, as seen from the southern hemisphere. As one astronomer phrased it, "the head was over Manly and the tail was over the Blue Mountains." For those unfamiliar with the geography of Sydney, where this astronomer was based, Manly is a beach-side suburb on the east coast and the Blue Mountains is the name of a plateau region to the west. The point of the description is well taken!

About the actual time of transit, the Earth must have passed through the comet's tail system, and we probably encountered a small amount of cometary matter, although not sufficient to give rise to any observable effects such as was noted in 1861 during the Earth's passage through the tail of Tebbutt's Comet. It is possible that bright moonlight and strong auroral activity (caused by active sunspots, *not* related to the comet) at the time of the passage may have masked minor sky illuminations, but it is unlikely that the rather conspicuous effects associated with the 1861 passage would have passed unnoticed, had they been repeated in 1910.

About the time the Earth passed the comet's tail, many observers reported seeing the tail rising out of both the evening and morning twilight as it extended out past the Earth.

Unfavorably changing perspective reduced the apparent length of the tail, even though the actual length continued to increase for a while after conjunction. At the end of May, the comet was a prominent evening object with a tail of 27 degrees and a nuclear magnitude of approximately 2. The total magnitude was somewhat higher than this, possibly still being about first magnitude (the maximum brightness of 0.7 magnitude having occurred just prior to inferior conjunction in mid-May).

As the comet moved away from both Earth and Sun, it faded and fell below naked-eye visibility during July. Nevertheless, it was traced with the aid of powerful telescopes until May 27, 1911, by which time its magnitude was about 15, a certain unsteadiness and tendency to fluctuate having been noted the previous month.

PROSPECTS FOR THE NEXT RETURN

At the end of 1979, P/Halley was located very close to the star Beta Canis Minoris, but with a magnitude of about 25 no one saw it.

Nevertheless, it draws nearer all the while, and the prospect of seeing it in only a few more years is an exciting one for many who have heard stories about its display in 1910. The question we may well ask, though, is "Will it live up to the reputation of its last return, will it be as spectacular in 1986 as it proved to be in 1910?"

To answer this, the most important tool needed is knowledge of the orbit to be followed by the comet in 1986 and of the relative comet-Earth geometry. Fortunately, both are known already with a high degree of accuracy, and we can at least put forward some fairly reasonable suggestions concerning its next return.

First, the comet will come to perihelion in the first half of the year (as it did in 1910), implying that it will be more favorably placed after perihelion than before. This is a characteristic of P/Halley and was also noted in 1910. In fact, the date of perihelion will be even earlier in the year—on February 9.

Secondly (and again similar to 1910), the comet will be most favorably placed from the southern hemisphere (see Figure 13).

However, the similarities between the two returns are now just about exhausted. The close approach to Earth near the inferior conjunction of mid-May 1910, and the consequent high brightness and very long apparent tail length, will *not* be repeated in 1986. This point needs some stress, as already there is a rumor circulating to the effect that the comet will come much closer, with the possibility of a direct collision with the Earth. Let me stress that there is *no possibility* of a collision between the Earth and Halley's Comet in 1986. At its closest approach, the comet will be 0.42 A.U. distant (April 1986), and even then the comet will be beyond the orbit of the Earth. There is no chance that we will pass through the tail of the comet next time either—the comet-Earth-Sun configuration precludes all possibility of a repeat of the 1910 event.

With these thoughts in mind, let us look at some of the more positive aspects of the 1986 return. Basing our discussion upon the orbital calculations of D. K. Yeomans[8] and the magnitude predictions of B. G. Marsden and R. G. Roosen,[9] we may expect the comet to become accessible to well-equipped amateurs during August 1985, when it will be visible in the eastern sky before daylight at approximately magnitude 14. Yet well before this date, professional astronomers will certainly be observing the comet. Indeed, searches have been conducted at every opposition since 1977 (using the largest telescopes), and it is probable that these will lead to the recovery of the comet in the first two years

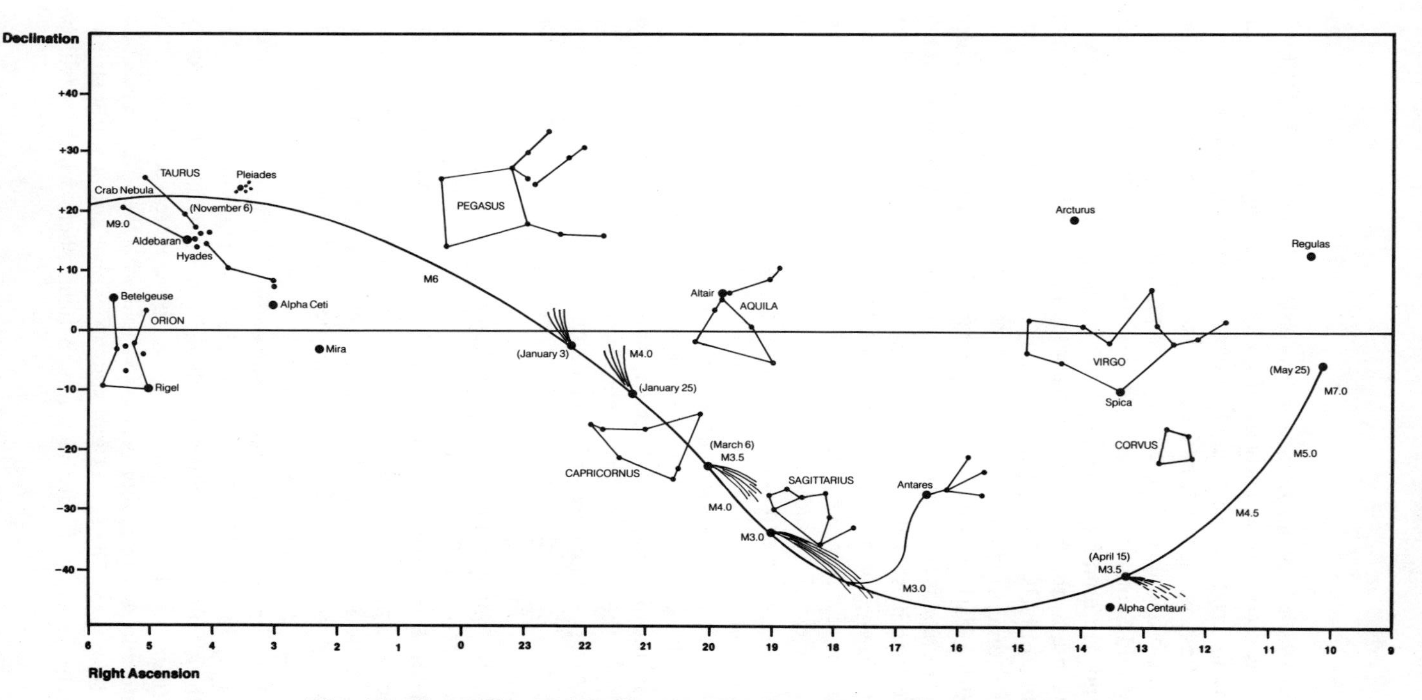

Fig. 13. *The path of P/Halley for 1985–86. Approximate tail length and orientation has been included, together with predicted magnitudes.*

of the 1980s. The magnitude will probably be about 23–24 if the comet is recovered as early as this, and the comet will steadily increase in brightness by about four or five magnitudes each year until 1984, when it will be close to the star Xi Geminorum. After this, the brightness will increase more rapidly.

By mid-September 1985 P/Halley should be about magnitude 12 and one month later, magnitude 10. Late in October, it passes very close to the famous Crab Nebula, and since the brightness of the two objects will probably be similar, some interesting comparisons should be possible.

During early November, the comet will continue to brighten and will be easy to locate near the Pleiades. At opposition in mid-November it is likely to be about magnitude 8, but will brighten more slowly during the remainder of the month as its decreasing heliocentric distance is offset by its increasing geocentric distance. Nevertheless, first indications of tail formation are likely to be noted toward the end of November or early December.

The comet is unlikely to become a naked-eye object during December, although it may reach about sixth magnitude late in the month. Probably the first naked-eye observations will be made late in the first week of January, and by January 13 the comet will join the crescent Moon and Jupiter in the western evening sky. At approximately fifth magnitude and with a tail of perhaps 5 degrees in length, the comet may by then have become an impressive sight for binocular observers.

Unfortunately, the comet's movement into twilight and the waxing Moon will make observations increasingly difficult in the weeks ahead. In fact, by the time of the full Moon (when the comet will possibly be about fourth magnitude), it will be already lost in the twilight. By the time of perihelion, the comet will be very close to the Sun in the sky and beyond it—an impossible position for earthbound observers.

After perihelion, the comet will emerge into the morning sky and may initially be about magnitude 2.5 or 3.0. Moving rapidly south, it will have attained an elongation of some 20 degrees by February 19. The tail should by then have increased somewhat in length and will probably become visible with the naked eye about this time also. The comet will start to evolve into a fairly striking object, especially from the southern hemisphere where it will be visible high against a dark sky throughout March and April. The brightness may increase again during March, if the 1910 trend is followed, reaching a second maximum of about third magnitude at the end of the month. This, it must be admitted, is not especially brilliant, although it is quite considerable for a comet. However, the high altitude of the comet (for southern hemisphere observers) and its appreciable tail length, should render it a spectacular object for favorably placed observers.

The tendency to grow long tails seems a fairly constant characteristic of this comet, and if this trend is again followed, a tail length of at least 20 degrees may be expected in late March and early April.

The magnitude attained in late March is likely to be approximately held during early April as the comet approaches the Earth —passing it on April 10.

After passing us by in April, the comet will move north into the evening sky and be located in Crater (between Corvus and Leo) by the month's end. At this time, northern hemisphere observers may have their best view of the comet—as an object of about magnitude 5 with a 1-degree tail.

During May, the comet may be expected to be about magnitude 7, fading to 11 in July as it becomes lost in the evening twilight. Further observations will probably be made late in 1986 and early 1987, when the magnitude of the comet is likely to be within the range of 14–15. It is even possible that still further observations may be obtained during the 1987–88 opposition, although the comet may, by then, be little brighter than magnitude 21 or 22.

As may be appreciated from reading this brief account of the comet's approaching return, it is clear that comet watchers of our own generation will not be greeted by the spectacle which thrilled, and frightened, many of our forefathers.

Yet, if all goes well, the coming return of Halley's Comet should provide us with a spectacle unequalled by any previous comet—namely, a really close look at an active cometary nucleus. This, at least, is the hope of European and Soviet astronomers, who plan to send space probes to the comet in an attempt to further unravel the mystery of the nucleus.

Present plans call for a Soviet-French flyby, a more sophisticated intercept mission proposed by the European Space Agency (ESA), and possibly a Japanese mission. The latter, however, is not destined to pass close to the comet, and for any really close views of the inner regions of the coma, we will need to rely upon the first two probes.

In actual fact, the ESA probe (named "Giotto," in honor of the comet's portrait painter) is likely to provide the most detailed information, as present data suggests that the Franco-Russian probe will not be equipped with a dust shield to protect it from the cosmic hailstorm of the inner coma. If this probe ventures too close to the nucleus, it will be a suicide mission, rather similar to the early lunar probes which were destined to end their lives ignominiously in a puff of moondust.

Giotto, on the other hand, will be a sophisticated 60-kg package of scientific instruments, and the Europeans are optimistic about its performance. It has two major shortcomings, however—it will lack stabilization (it will be a spinning spacecraft), and its targeting accuracy will not be as great as the best American spacecraft (about 500 km for Giotto, as compared to less than 90 km for a U.S. craft).

In view of the greater sophistication of U.S. space technology, comet specialists would be truly delighted if that nation would also contribute to the celestial armada bound for Halley but, alas, present indications are not encouraging. Indeed, the last few years have seen a winding down of the U.S. scientific space program, and this, more than anywhere else, has been reflected in the diminishing proposals for a U.S. Halley probe.

Originally, an optimistic plan was proposed involving a launch in 1981 and a rendezvous with the comet in deep space. The spacecraft was then supposed to pace the comet as the latter moved toward perihelion, possibly even landing on the nucleus!

Needless to say, this plan soon went to wherever ambitious proposals go when they die, only to be succeeded by one almost as extravagant. This new plan, put forward by NASA's Comet Science Working Group (CSWG), involved an intercept of Halley prior to perihelion (during which a module would be dropped into the inner coma) before continuing on to a rendezvous with Comet Tempel 2 in 1988. One proposal involved the probe's actually going into orbit about the nucleus of this latter comet.

Unfortunately, the success of this endeavor was contingent upon the development of the ion drive and the Space Shuttle; and when the former was dropped and the latter fell behind schedule, this second Halley mission followed the first into the depths of limbo.

The most recent U.S. plan calls for a conventional spacecraft, launched from either the Shuttle or a Titan rocket, to fly within 1,000 km of the nucleus of Halley. The 125-kg scientific payload would consist of nine experiments—namely, a neutral mass spectrometer, ion mass spectrometer, electron analyzer, magnetometer, plasma-wave analyzer, dust-composition analyzer, dust counter, remote sensor, and cameras. At its closest approach, details as small as twenty meters would be visible on the nucleus, and when far from the comet, the probe would enter an observatory phase during which the head and tail of Halley could be scrutinized in a manner not possible for earthbound observers.

The probe would return to near the Earth late in 1986 and would use the gravity-assisted trajectory changes to send it on to Encke's Comet (September 1987), or Borrelly's Comet (January 1988), or a double encounter with the Apollo-group asteroid Geographos in September 1987 and Comet Tempel 2 exactly one year later.

Alas, at the time of writing, prospects for the funding of such an interesting and valuable mission are not encouraging—and time is rapidly running out! If funds have not been allocated by the time you read these words, the launch date for a U.S. Halley mission will be considerably postponed—until late 2060, in time for the comet's next return in July 2061. Let us hope that funds will be available then!

IS HALLEY'S COMET FADING?

Almost inevitably, the discussion of a periodic comet raises the question of secular fading. Already, in Chapter 2, we have discussed this issue, and I do not wish to repeat the general arguments again here; but nevertheless, the specific issue of the fading of P/Halley cannot be simply ignored—especially in view of the fact that the relatively faint return predicted for 1986 may well be used by protagonists of this position to support their theories. Of course, the predicted magnitudes at the comet's next return are expected to be relatively low because of the object's greater distance rather than any supposed fading—if it is true that secular fading is taking place, the comet may be even fainter than we have supposed.

The chief protagonist is, of course, Vsekhsvyatsky, whose speculations concerning P/Halley form part of his general thesis concerning periodic comets. Vsekhsvyatsky's estimations of the absolute magnitudes of P/Halley at various returns are as follows:[10]

Year	*Absolute Magnitude*
760	2.0
837	2.0
912	1.7
989	3.5
1066	–1.0 or 0.0
1145	2.0
1222	?
1301	3.0
1378	3.5
1456	4.2
1531	4.0
1607	3.5
1682	4.0
1759	3.8
1835	4.4
1910	4.6

Some recent speculations give an absolute magnitude of about 5.6 for the forthcoming 1986 return,[11] in line with this general position.*

* This value, by Konopleva and Shul'man, is based upon their theory of the evolution of the nucleus and not upon alleged secular fading. Nevertheless, their predicted value for 1986 is of interest here as it seems to imply a prior acceptance of the fading hypothesis. Thus the fact that the derivation of a lower value for the intrinsic brightness of

The first thing which is conspicuous in the above list is the discordant value for 1066. If the above is to be believed, the comet's absolute brightness on this return—*and on this return alone*—was among the highest ever recorded for a comet. Unless we are to speculate about anomalous outbursts or the like, this tends to look just a little suspect.

Secondly, translating old observations into the modern magnitude scale is always bound to be treacherous. A comet is not like a star, and unless one has some grounding in astronomy and in the practice of making magnitude estimates, it is very easy to grossly overestimate the brightness of a large and extended object. For instance, nonastronomical people have said that the brightness of P/Halley in 1910 was "between that of Venus and the full Moon," whereas in actual fact it was never appreciably above first magnitude. Nevertheless, the large comet with its long tail was as *conspicuous* as an object like the full Moon, and it is easy to translate this in terms of brightness. We must expect that many early reports of comets—where this factor is also mixed with fear—would show similar exaggeration, especially in view of the fact that old reports often use phrases such as "as *large* as Jupiter" or even "as large as the Moon," and the exact meaning of "large" in this context is not at all clear.

This is not the only difficulty, however. The absolute magnitude values listed previously have all been reduced to a value derived on the assumption of a value of n equal to 4, which Vsekhsvyatsky believes to be the average for all comets. Nevertheless, is this value really representative of *P/Halley?* If it is not, the derived values will not be accurate. Indeed, the comet does appear to increase in brightness more rapidly than this value of n indicates, and this fact must be considered when deriving the comet's photometric parameters.

Similarly, all such assumptions imply a symmetrical light curve relative to perihelion. As a matter of fact, comets are frequently not as well behaved as this,[12] and a number of periodic objects (most notably P/d'Arrest[13]) show large-scale asymmetries as a

---

the comet in 1986 is accepted as a prediction and not as a difficulty for their theory would seem to indicate favorable disposition toward the hypothesis that the comet is fading. (Of course, if they are correct, the comet will be a fainter visual object than assumed in our above discussion.)

regular aspect of their light curves on every well-observed return. Any such feature in the light curve of P/Halley could also result in major miscalculations of its intrinsic brightness.

In fact, it does appear that just such an asymmetry exists for P/Halley. At its most recent return, the comet increased in brightness rather more rapidly than average, having a value of n near 5, but after perihelion the value of n fell considerably and seemed, in fact, to actually decrease as the comet's heliocentric distance increased. Thus the comet faded at a decreasing rate and by July 1910 was some three magnitudes brighter than expected on the basis of pre-perihelic predictions.[14] This trend also seemed apparent during the comet's previous return. Thus on January 25, 1836, the comet appeared to the naked eye as a bright (though tailless) object of almost second magnitude, even though its heliocentric distance had by this time increased to 1.48 A.U. and its geocentric distance to 1.60 A.U.[15] If the intrinsic brightness of the comet at this date was really representative of the absolute magnitude during the entire return, the comet would have achieved negative magnitudes in mid-October 1835 when peak brilliance was attained. In fact, observations during this month indicate that the comet was at no time brighter than first magnitude.

It is indeed very probable that a similar trend is followed by this comet at every return. Attempts to fit the long period of naked-eye visibility associated with some appearances of this comet coupled with vivid descriptions of the comet which probably reflected the observer's reaction rather than the object's magnitude, have (I would suggest) been responsible for the very high absolute magnitude estimates at some of the early returns—in particular that of 1066. It may also be worth noting that the comet may be prone to fluctuations in brightness—flares were certainly noted in late 1909—and if this is a constant property of the comet, we have yet another factor complicating the determination of a light curve. Almost certainly astronomers will be on the watch for any such activity in 1986.

Thus with all these "ifs," "buts," and sundry caveats in mind, I think it is rather premature to be too dogmatic about the secular behavior of this comet, but we can, I believe, at least cast doubt upon the main evidence presented in favor of a *substantial* fading over the last millennium. Of course, the comet, like all other active comets, is slowly wasting as material is evaporated from the nucleus into the coma and tail, but the resources of P/Halley are probably quite large, and the amount lost to the comet at each return is only a tiny proportion of its total reserves. We may, therefore, expect the comet to continue producing impressive displays for many returns yet to come, unless the unpredictable happens and, like the comets of Biela and Westphal, Halley's suddenly and unexpectedly fades from our sight. Let us hope that this is a contingency which does not eventuate!

REFERENCES

1. M. Kamienski. "Halley's Comet in the Time of Hammurabi"; "Unknown Comet About 2008 B.C. Was Probably Halley's." *Journal of the British Astronomical Association.* Vol. 70 No. 7, Session 1959–60, pp. 304–17.

2. D. Justin Schove. "Halley's Comet." *Journal of the British Astronomical Association.* Vol. 65 No. 7, July 1955, pp. 285–89.

3. 1 Chronicles 21.

4. D. Justin Schove. "The Comet of David and Halley's Comet." *Journal of the British Astronomical Association.* Vol. 65 No. 7, July 1955, pp. 289–90.

5. B. G. Marsden. *Catalogue of Cometary Orbits* (Third edition). Cambridge, Mass.: Smithsonian Astrophysical Observatory, 1979.

6. T. Kiang. "The Past Orbit Of Halley's Comet." *Memoirs of the Royal Astronomical Society.* Vol. 76, 1972, p. 27.

7. R. J. M. Olson. "Giotto's Portrait of Halley's Comet." *Scientific American.* Vol. 240 No. 3, May 1979, pp. 134–42.

8. D. K. Yeomans. "Comet Halley—the Orbital Motion." *Astronomical Journal.* Vol. 82, 1977, pp. 435–40.

9. R. G. Roosen and B. G. Marsden. "Observing Prospects for Halley's Comet." *Sky and Telescope.* Vol. 49 No. 6, June 1975, pp. 363–64.

10. S. K. Vsekhsvyatsky. "Physical Characteristics of Comets." Israel Program for Scientific Translations, p. 44.

11. V. P. Konopleva and L. M. Shul'man. "On the Sizes of

Cometary Nuclei." In *The Motion, Evolution of Orbits, and Origin of Comets,* p. 282.

12. Luigi G. Jacchia. "The Brightness of Comets." *Sky and Telescope.* Vol. 47 No. 4, April 1974, pp. 216–20.

13. J. E. Bortle. "The 1976 Apparition of Periodic Comet d'Arrest." *Sky and Telescope.* Vol. 53 No. 2, Feb. 1977, pp. 152–56.

14. S. K. Vsekhsvyatsky. "Physical Characteristics of Comets," p. 163.

15. Luigi G. Jacchia. "The Brightness of Comets."

APPENDIX ONE
SUGGESTED OBSERVATIONS

COMA DIAMETER

Does the real diameter of a comet's coma change as it approaches and recedes from the Sun?

This is still a largely unsettled question, the answer to which may well vary from comet to comet—with some objects contracting and others expanding as they approach the Sun.

It may seem an easy matter to settle this question, but alas (as with so many issues concerning comets) things are not as straightforward as they might, at first sight, appear.

For instance, we are faced with the apparent paradox that higher magnifications tend to give *smaller* coma diameters unless the comet is a very small and condensed one. The paradox is resolved, however, if the diffuse nature of the coma is kept in mind —the greater the power of the eyepiece, the more an equal quantity of light will be "spread," so to speak, resulting in a diminution of the intensity of the image to a point where the outer fringes of the image are so reduced in intensity that the human eye is not sufficiently sensitive to detect them against the field of the telescope. For a diffuse image such as the average comet, there is no clear-cut boundary; and contrast, therefore, is the primary factor. This contrast will be greatest when the image is small and concentrated in a wide telescopic field—a situation best achieved with low-power eyepieces and short-focus telescopes or, especially, high-power binoculars such as 15 × 80, 20 × 80 or 20 × 65.

Magnification is not the only factor influencing estimates of coma diameters. Sky illumination (either natural or artificial) and haze are other important factors, especially as a comet moving toward perihelion is often moving into twilight and encountering increasing sky illumination. As the fainter, outer regions of the coma of such a comet become overpowered by the increasing twilight, an observer experiences the illusion of a shrinking coma, even though a hypothetical observer in outer space (where the twilight scattering of sunlight would not be a problem) may see no such thing.

Similarly, artificial lights and city haze will reduce the apparent observed diameter of a comet, and this may be confusing in cases where (say) a city lies to the west of an observer and an open area of country lies to his or her east. In this case, a morning comet moving toward perihelion will be seen in a much darker sky than the same object reappearing in the west after perihelion, and this would almost certainly result in smaller diameter estimates after perihelion. All such contingencies due to local environment must be taken into consideration when observations of this kind are being made, in addition to the more obvious factors such as moonlight, airglow, auroras, and the like.

Measurements of coma diameters may be made directly by means of a reticle-equipped eyepiece or filar micrometer or, where these instruments are not available, by comparison of the coma with separations between stars seen in the telescopic field (as listed in a star atlas, preferably one of large scale).

Alternately, for comets having declinations less than ±70 degrees, the diameter may be ascertained by the so-called "drift method." That is to say, an eyepiece with crosshairs (one of which is north-south oriented) is used, and the comet is allowed to drift across the field of the eyepiece; the length of time (in seconds) between its first and last contact with the crosshair is then measured. After this is performed several times and an average is taken, the coma diameter may be calculated by using the formula

$$d = 0.25T \cos \delta$$

where d is the coma diameter in minutes of arc, T is the length of time (in seconds) between the comet's first and last contact with the crosshair, and δ is the comet's declination.

From the apparent diameter, the "reduced diameter" (d') may be calculated

$$d' = d\Delta$$

where Δ is the comet's geocentric distance in astronomical units. Changes in the size of d' (given in minutes of arc) enable the intrinsic expansion or contraction of the coma to be noted.

These values may be converted into kilometers by noting that a sphere of 43,517 kilometers at a distance of 1 A.U. subtends an angle of 1 minute of arc. Therefore, the coma's diameter in kilometers (D) can be found by

$$D = 43{,}517d' \qquad (\text{or } D = 43{,}517d\Delta)$$

Observations of the coma should also include, in addition to the diameter, an estimate of the degree of condensation. This is given on a scale ranging from 0 (a totally diffuse comet of uniform surface intensity—i.e., having no increase in intensity toward the center of the coma) to 9 (a comet which appears like a stellar point or like a small planetary disk).

LENGTH OF TAIL

The tails of comets vary in length as the comet approaches and recedes from the Sun and sometimes may appear to be sensitive to solar activity as well. Continuous measurement of the tail is therefore important, as it is only by means of a good series of observations that such variations can be discovered (as, for instance, the relationship $1/r^2$, where r is the comet's heliocentric distance, for the tail of Comet Arend-Roland of 1957).

Estimation of the apparent length of short tails is fairly straightforward. Simple plotting of the position of the comet and the end of its tail on a star atlas and direct measurement of this is sufficient. Indeed, for very short tails it is possible to employ the same method as the second of those mentioned above for estimating coma diameters—namely, comparison with the known separation of field stars. Even estimates comparing the tail with the diameter of the eyepiece field should be sufficient for very short tails.

When the tail is very long (greater than 10 degrees), however, scale distortions inherent in map projections will render direct measurement from a star atlas too inaccurate to be reliable. For such tails, the following formula must be employed

$$\cos l = \sin \delta_1 \sin \delta_2 + \cos \delta_1 \cos \delta_2 \cos (a_1 - a_2)$$

where l is the apparent tail length in degrees, a_1 *and* δ_1 are the right ascension and declination of the comet's head, and a_2 and δ_2 are the right ascension and declination of the end point of the comet's tail, also given in degrees.

In order to reduce these angular values to "real" values, the following formulae may be used

$$L = \Delta \frac{\sin l}{\sin (\alpha - l)}$$

where L is the length of the tail in astronomical units, l the apparent length of the tail in degrees, and α is the phase angle, the latter being given by

$$\cos \alpha = \frac{r^2 + \Delta^2 - R^2}{2r\Delta}$$

OR

$$\tan \frac{2\alpha}{2} = \frac{(p - r)(p - \Delta)}{p(p - R)}$$

OR

$$\sin \tfrac{1}{2}\alpha = \tfrac{1}{2} \sqrt{\frac{(R - r + \Delta)(R + r - \Delta)}{r\Delta}}$$

where $p = \frac{1}{2}\,(r + \Delta + R)$ and R is the radius vector (heliocentric distance) of the Earth.

Alternatively, if the elongation, rather than the phase angle, is used, the tail may be calculated by

$$L = \frac{r\,\Delta \sin l}{R \sin (E + l) - (\Delta \sin l)}$$

where E is the comet's elongation and may be calculated by this formula:

$$\cos E = \frac{(\Delta^2 + R^2 - r^2)}{2\Delta R}$$

In addition to estimates of the length of a comet's tail, the position angle (*p.a.*) should also be given. This is the angle in which the tail points and is given in degrees (0° = north, 90° = east, etc.), and is most easily measured by plotting the tail on an atlas and measuring the angle with a protractor.

PHOTOMETRIC ESTIMATES

There are four methods by which the total brightness (m_1) of a comet may be estimated, namely:

a. *The Bobrovnikoff Method.*[1] Comparisons are made between extrafocal images of the comet and comparison stars of nearly equal brightness. Two or, preferably, three stars are used, one being a little brighter and another a little fainter than the comet. Using a low-power eyepiece (about 1.4–2 magnifications per centimeter of aperture, which makes 15 × 80 binoculars just about perfect for the brighter comets), the instrument is defocused until the star images and the comet have about the same diameter. Direct comparisons can then be made—e.g., if the comet's brightness appears to be halfway between a star of magnitude 6 and another of magnitude 7, the estimate will be magnitude 6.5.

b. *The Sidgwick (or "In-Out") Method.*[2] This method involves the comparison of the in-focus image of the comet with extrafocal images of comparison stars. Unless binoculars with individual focus mounts are employed, this involves the memorizing of the image of the comet for comparison with the out-of-focus images of the comparison stars and requires rather more skill and experience than *a* above.

c. *The Beyer Method.*[3] In this method, the eyepiece is racked out of focus until the images of the comet and comparison stars become so expanded and diffuse as to be rendered invisible. The comparison is made by noting which vanishes first. This method is not generally used as it requires considerable skill and is very sensitive to sky illumination.

d. *The Morris Method.*[4] The image of the comet is put out of focus sufficiently to obtain an approximately uniform surface brightness. The size and surface brightness of this image is memorized (unless binoculars with individual focus mounts are being used), and the comparison stars are put out of focus until they are the same size as the out-of-focus comet image. The estimate is made by comparing the surface brightness of the out-of-focus comparison stars with that of the out-of-focus comet, and the process is repeated until a satisfactory estimate is obtained. (For a small comet, this method approaches *a,* whereas for a large and very diffuse one, it becomes *b,* above.)

All these methods are subject to certain systematic errors which depend upon a variety of factors—e.g., the sensitivity of the observer's eyes to low levels of illumination, whether the telescope gives a dark field, the presence of slight haze or enhanced airglow, twilight, moonlight, glow from artificial lighting and the appearance of the comet itself (for instance, a very large and diffuse object is not easily compared with a point source of light, however far the images are put out of focus), and any mismatch between the color of the comet and that of the comparison star. Regarding this latter point, it is advisable to stay away from red stars when making comparisons. If possible, only stars of spectral type G or hotter should be used as comparison stars (i.e., only stars which are yellow, white, or blue).

In general, the Bobrovnikoff method is the easiest, although a certain unavoidable difference in the size of the out-of-focus images of comparison stars and the comet will be a potential source of error. Also, the method becomes difficult when large comets (such as those passing very near the Earth) are concerned, especially if one is using binoculars which can only be racked out of focus to a limited extent. Again, it may not be possible to use this method for very faint diffuse comets near the limit of visibility. For these latter objects, the Sidgwick method is more suitable. If the comet is not only large and diffuse, but possessed of a sharp central condensation and a much fainter outer coma, the Morris method will be the preferred one.

Magnitudes of the comparison stars will be supplied by such catalogues as the Smithsonian Observatory Catalogue (although only those stars coded "H" or "T" should be used for accurate comparisons), the Yale Bright Star Catalogue, or the variable star sequences issued by the American Association of Variable Star Observers. Listed magnitudes of nebulous objects and globular star clusters or the magnitudes given in star atlases should *never* be used for this purpose.

In order to compute the absolute magnitude (m_0) and "activity parameter" (n) of a comet according to the usual formula

$$m = m_0 + 5 \log \Delta + 2.5 n \log r$$

the magnitude values are first reduced to a standard distance of 1 A.U. (by removing $5 \log \Delta$) and to a standard telescope aperture of 6.8 cm, by means of this formula

$$m' = m - c\,(A - 6.8)$$

where m' is the corrected magnitude, m is the observed magnitude, A is the telescope's aperture in centimeters, and where the value of c differs for reflectors and refractors as follows:

Reflectors:	$c = 0.019$
Refractors:	$c = 0.066$

These reduced values (like the reduced values of tail length and coma diameter) will be of more concern to directors of

Comet Sections and professional astronomers than to the lone amateur observer. The latter will be mainly concerned with obtaining the "raw" observations and sending them to such bodies as the IAU, the *International Comet Quarterly,* or to the Comet Sections of the various astronomical societies.

ACTIVITY WITHIN THE COMET

A careful watch by amateurs is often responsible for the discovery of any unusual activity within the coma or tail. For instance, a comet's brightness may dramatically increase within a matter of hours, or jets and envelopes may form with equal rapidity within the coma. Accurate observations of such activity—including the exact time when it was initially detected, plus measurements of position angles of jets, diameters of halos and envelopes, and position angles of secondary condensations and the like—can prove especially valuable and may be all that is required by a professional astronomer to enable him to compute the rotation of the comet's nucleus.

It is advisable to search the inner condensation of a comet several days after an abrupt flare has been noted, using the highest magnification possible, as sometimes (e.g., comets Tago-Sato-Kosaka in 1969 and West in 1976) such a flare is a prelude to the breaking up of the comet's nucleus; and early reports of such a phenomenon, together with accurate times of the observations (in Universal Time, preferably given to the decimal of a day) and position angles of secondary nuclei, are always welcomed by professional comet specialists.

REFERENCES

1. N. T. Bobrovnikoff. *Contributions of the Perkins Observatory*. Bulletin Nos. 15 and 16, 1941.
2. J. Sidgwick. *Observational Astronomy for Amateurs*. London: Faber and Faber, p. 251.

3. M. Beyer. "La Physique des Comêtes." *Mémoires de la Société Royale des Sciénces de Liege.* Vol. 13, 1952, p. 236.

4. C. S. Morris. *Comet News Service.* Vol. 79 No. 1, 1979, pp. 2–3.

APPENDIX TWO
CENTURIES OF COMETS

The 1979 edition of Marsden's Catalogue of Cometary Orbits (obtainable from the Central Bureau for Astronomical Telegrams of the International Astronomical Union) lists the orbits of 1,027 comets (including returns) up to the end of 1978. Between that date and October 1979, a further nine orbits had been calculated (for comets appearing in 1979). The number of objects for which records exist is, however, much greater. A general list up to the same time contains something like 2,153 objects.

Needless to say, many of these objects are doubtful. Early records did not clearly distinguish between comets and other astronomical (and perhaps even meteorological) phenomena, and it is probable that lists of comets also include such phenomena as meteors, novas and supernovas, auroras, and perhaps even odd clouds and distant lightning. (One phenomenon which may have found its way into such lists is the comparatively rare display of a moonlight rainbow. I was fortunate in seeing one of these several years ago and was struck by its similarity with the tail of a bright comet. This is especially apparent when the rainbow is only partially formed and the colors are not distinct. Is there any tendency for one-report "comets" to be around the times of the full Moon? Like real comets, "moonbows" tend to be found in the west after sunset and the east before dawn but at times near the full Moon—i.e., they are located opposite the full Moon in the sky.)

Sometimes a single comet may even be reported as two (or even, on occasions, up to four) separate comets. Objects of high declination—visible after sunset and again before dawn—are sometimes recorded as two separate objects, as are comets which are bright both before and after conjunction with the Sun.

Spurious comets or "objects that get away" are not confined to the past. In fact, in the period mid-1978–mid-1979, there were five unconfirmed comet discoveries—all photographic. Three of these were almost certainly faint comets which were recognized on photographs exposed some months previously—too late to allow recovery and confirmation. One object may have been a minor planet as the image was only marginally diffuse, and the fifth object appears to have been a misinterpretation of a photographic defect.

On the other hand, many comets are known for which orbital elements have not been obtained. Most of these are old objects, known only in early records, and the positional descriptions are simply not sufficiently accurate to enable determination of their orbits. This is not surprising—what *is* surprising is that astronomers have been able to compute as many orbits for early objects as their catalogues do in fact contain!

The following is only intended to give a *rough* idea of the number of comets (including returns) recorded by centuries. No attempt has been made to sift real comets from spurious ones or even legendary ones. Such a sifting can never be completely accurate, and all I have attempted to present here is the raw data from previously compiled lists (principally those of Chambers, Baldet, Ho Ping-Yu, and Vsekhsvyatsky.

Century	*No. of Comets*
All B.C.	155
1st A.D.	31
2nd	42
3rd	45
4th	40
5th	45
6th	46
7th	41
8th	34
9th	57
10th	60
11th	71
12th	62
13th	59
14th	62
15th	80
16th	83
17th	41
18th	80
19th	364
20th*	655

* (To Oct. 1979)

APPENDIX THREE
SOME INTERESTING COMETS

The following three tables concern comets which, for one reason or another, are especially interesting.

Table 1 concerns those comets which have periods of less than 200 years and have been observed at more than one perihelion passage. The first column gives the comet by name, the second the latest perihelion passage (prior to October 1979) at which it was observed, P is the period in years, q the perihelion distance (heliocentric) in astronomical units, and H_0 is the approximate absolute magnitude at the latest return—i.e., the hypothetical magnitude of the comet at a distance of one astronomical unit from both Earth and Sun. In general, this is based upon the assumption that the comet's brightness according to $m = H_0 + 5 \log \Delta + 2.5n \log r$ where Δ and r denote the comet's geocentric and heliocentric distance, respectively. The quantity n is usually assumed to equal 4 (i.e., $2.5n = 10$), and where this assumption is not made, the value of $2.5n$ (*NOT* merely n) is included in parenthesis. Most magnitude estimates are assumed to refer to the total or globular magnitude (m_1) although where the estimates obviously refer only to the nuclear brightness "m_2" is included parenthetically.

Table 2 includes those comets which have periods of less than 200 years, but which have only been observed at a single return.

Both tables have some notable omissions. Thus most lists of

comets seen more than once include P/Grigg-Mellish of 1907 ($P = 164.3$ years) assumed to have been the return of the comet of 1742. However, this comet has now been shown to have a period of hundreds of years at least and can no longer be considered periodic (the 1742 comet was obviously a different object).

Table 2 omits comets Ross of 1883 (the 63-year period sometimes given being completely fictitious), Wilson-Harrington of 1949 (the period of 2.3 years being uncertain), and the doubtful objects Kulin of 1939 ($P = 5.6$ years) and Perrine of 1916 ($P = 16$ years).

A note on lost comets follows and a table of daylight comets completes the Appendix.

Table One

Periodic Comets of More Than One Apparition

Comet	Latest Perihelion Date (prior to Oct. 1979)	P	q (A.U.)	H_0	Remarks
Encke	1977 Aug. 16	3.31	0.34	11.0	
Grigg-Skjellerup	1977 Apr. 11	5.10	0.99	12.5	Brightness increase after perihelion at this return; very close (0.18 A.U.) to Earth in April.
Tempel 2	1978 Feb. 20	5.28	1.37	10.0	
Honda-Mrkos-Pajdusakova	1974 Dec. 28	5.28	0.58	13.4 (19.1)	
Schwassmann-Wachmann 3	1979 Sept. 2	5.32	0.94	11.5	Lost since 1930; recovered 1979.
Neujmin 2	1927 Jan. 16	5.43	1.34	11.3	
Brorsen	1879 Mar. 31	5.46	0.59	9.2	Lost.
Tempel 1	1978 Jan. 11	5.50	1.50	9.4	
Clark	1978 Nov. 26	5.51	1.56	12.0	
Tuttle-Giacobini-Kresak	1978 Dec. 25	5.58	1.12	13.0	Fluctuations in brightness.

Table One

Periodic Comets of More Than One Apparition

Comet	Latest Perihelion Date (prior to Oct. 1979)	P	q (A.U.)	H_0	Remarks
Tempel-Swift	1908 Oct. 5	5.68	1.15	12.8	
Wirtanen	1974 July 5	5.87	1.26	16.3 (m_2)	
d'Arrest	1976 Aug. 12	6.23	1.16	2.8 (110) preperihelion 6.5 (13.2) postperihelion	Peculiar light curve; close approach to Earth.
Du Toit-Neujmin-Delporte	1970 Mar. 21	6.31	1.68	14.0	
De Vico-Swift	1965 Aug. 23	6.31	1.62	14.5	
Pons-Winnecke	1976 Nov. 28	6.36	1.25	14.5 (m_2)	
Forbes	1974 May 20	6.40	1.53	10.0	
Kopff	1977 Mar. 7	6.43	1.57	13.4	
Schwassmann-Wachmann 2	1974 Sept. 12	6.51	2.14	11.4	
Giacobini-Zinner	1979 Feb. 12	6.52	0.99	10.0	Comet associated with Draconid meteor shower.

Wolf-Harrington	1978 Mar. 15	6.55	1.61	13.0	
Churyumov-Gerasimenko	1976 Apr. 7	6.59	1.30	10.0	
Biela	1852 Sept. 23	6.62	0.86	8.1	Lost; associated with Andromedid meteor shower.
Tsuchinshan 1	1978 May 10	6.66	1.50	14.0	
Perrine-Mrkos	1968 Nov. 1	6.72	1.27	18.5	Believed lost from 1909–1955; probably lost again.
Reinmuth 2	1974 May 8	6.74	1.94	10.5 (15.0)	
Borrelly	1974 May 12	6.76	1.32	13.0	
Johnson	1977 Jan. 8	6.76	2.20	10.0	
Gunn	1976 Feb. 10	6.80	2.44	10.0	
Harrington	1960 June 28	6.80	1.58	14.8	
Tsuchinshan 2	1978 Sept. 20	6.83	1.78	14.0	
Arend-Rigaux	1978 Feb. 2	6.83	1.44	8.9 (21.5)	Weak coma and tail.
Brooks 2	1974 Jan. 4	6.88	1.84	13.5	
Finlay	1974 July 3	6.95	1.10	13.0 (15.0)	

Table One

Periodic Comets of More Than One Apparition

Comet	Latest Perihelion Date (prior to Oct. 1979)	P	q (A.U.)	H_0	Remarks
Taylor	1977 Jan. 12	6.97	1.95	12.0	Seen previously in 1916 when it split; fainter nucleus only recovered—evidently this was the more massive of the two.
Holmes	1979 Feb. 22	7.06	2.16	13.5	
Daniel	1978 July 8	7.09	1.66	11.5 (15.0)	
Shajn-Schaldach	1979 Jan. 9	7.27	2.22	12.0	
Faye	1977 Feb. 27	7.39	1.61	9.5	
Ashbrook-Jackson	1978 Aug. 19	7.43	2.28	7.1	
Whipple	1978 Mar. 27	7.44	2.47	13.5	
Harrington-Abell	1976 Apr. 21	7.58	1.77	15.0 (m_2)	Passed 0.037 A.U. from Jupiter, April 12, 1974—a record for a comet

					seen both before and after such an approach.
Reinmuth 1	1973 Mar. 21	7.63	1.99	14.0	
Kojima	1978 May 24	7.85	2.40	10.0	
Oterma	1958 June 10	7.88	3.39	9.5	Orbit perturbed by Jupiter; comet not likely to be recovered.
Arend	1975 May 24	7.98	1.85	14.5	
Schaumasse	1960 Apr. 18	8.18	1.97	11.0	Outburst in 1952; badly placed since 1960.
Jackson-Neujmin	1978 Dec. 26	8.37	1.43	16.7 (m_2)	
Wolf	1976 Jan. 25	8.42	2.50	13.0	
Comas Sola	1978 Sept. 24	8.94	1.87	8.5 (15.0)	
Kearns-Kwee	1972 Nov. 28	9.01	2.23	11.2	
Denning-Fujikawa	1978 Oct. 20	9.01	0.78	11.5	Return of Denning 1 of 1881.
Swift-Gehrels	1972 Aug. 31	9.23	1.35	15.0	
Neujmin 3	1972 May 16	10.57	1.98	14.5	
Klemola	1976 Aug. 20	10.97	1.77	12.5	
Gale	1938 June 18	10.99	1.18	10.5	Favorable returns infrequent.

Table One

Periodic Comets of More Than One Apparition

Comet	Latest Perihelion Date (prior to Oct. 1979)	P	q (A.U.)	H_0	Remarks
Vaisala 1	1971 Sept. 12	11.28	1.87	13.5	
Slaughter-Burnham	1970 Apr. 13	11.62	2.54	13.6	
Van Biesbroeck	1978 Dec. 3	12.39	2.40	7.5	
Wild 1	1973 July 2	13.29	1.98	14.0	
Tuttle	1967 Oct. 28	13.77	1.02	9.0 (15.0)	
Du Toit 1	1974 Apr. 4	14.97	1.29	16.0	
Schwassmann-Wachmann 1	1974 Feb. 15	15.03	5.45	5.0	Visible throughout its orbit; brightness flares.
Neujmin 1	1966 Dec. 9	17.93	1.54	10.2 (12.5)	
Crommelin	1956 Oct. 25	27.89	0.74	10.7	
Tempel-Tuttle	1965 Apr. 30	32.91	0.98	13.5 (m_2)	Associated with Leonid meteor shower.
Stephan-Oterma	1942 Dec. 19	38.84	1.59	5.3 (28.0)	Sometimes called "Coggia-Stephan" in older lists.

Westphal	1913 Nov. 26	61.86	1.25	9.3 (—5.2)	Faded out in 1913.
Olbers	1956 June 19	69.47	1.18	5.5	
Pons-Brooks	1954 May 22	70.98	0.77	5.9 (11.7)	Brightness flares. One of the brighter periodical comets.
Brorsen-Metcalf	1919 Oct. 17	71.93	0.49	8.7 (12.6)	
Halley	1910 Apr. 20	76.09	0.59	4.6	Brightest periodical comet.
Herschel-Rigollet	1939 Aug. 9	154.90	0.75	8.5	Brightness fluctuation noted.

Table Two

Periodic Comets of Only One Appearance

Comet	Perihelion Date	P	q	H_0	Remarks
Helfenzrieder	1766 Apr. 27	4.51	0.40	6.8	Naked-eye tail.
Blanpain	1819 Nov. 20	5.10	0.89	8.5	
Du Toit 2	1945 Apr. 18	5.28	1.25	11.9	
Barnard 1	1884 Aug. 16	5.40	1.28	8.9	
Brooks 1	1886 June 7	5.60	1.33	8.9	
Lexell	1770 Aug. 14	5.60	0.64	7.7	Closest approach to Earth on record; orbit later altered by Jupiter.
Pigott	1783 Nov. 20	5.89	1.46	6.9	
Haneda-Campos	1978 Oct. 9	5.99	1.10	14.0	About the most favorable return possible; possible outburst at discovery; unsteady brightness and possible signs of exhaustion.
West-Kohoutek-Ikemura	1975 Feb. 25	6.11	1.40	10.0	
Russell	1979 May 26	6.13	1.61	15.0	

Wild 2	1978 June 15	6.17	1.49	6.5 (14.0)	
Kohoutek	1975 Jan. 17	6.23	1.57	12.0	
Tritton	1977 Oct. 28	6.33	1.44	16.5	
Spitaler	1890 Oct. 27	6.37	1.82	9.0	
Harrington-Wilson	1951 Oct. 30	6.38	1.67	12.1	
Kowal 2	1979 Jan. 13	6.51	1.52	14.5	
Barnard 3	1892 Oct. 30	6.52	1.43	9.8	First comet discovered by photography.
Giacobini	1896 Oct. 28	6.65	1.46	9.9	Faint satellite comet seen.
Schorr	1918 Sept. 30	6.66	1.88	11.0	
Giclas	1978 Nov. 21	6.68	1.73	13.5	
Longmore	1974 Nov. 4	6.98	2.40	11.5	
Swift	1895 Aug. 21	7.20	1.30	11.4	
Denning	1894 Feb. 9	7.42	1.15	10.4	
Schuster	1978 Jan. 6	7.48	1.63	13.2 (m_2)	
Metcalf	1906 Oct. 10	7.78	1.63	9.5	
Gehrels 2	1973 Dec. 1	7.94	2.35	11.0	
Gehrels 3	1977 Apr. 22	8.10	3.42	9.5	Visible throughout its orbit.
Smirnova-Chernykh	1975 Aug. 7	8.53	3.57	8.3	Visible throughout its orbit; orbital eccentricity less than that of Mars before approach to Jupiter in 1963 (e only 0.08 then).

Table Two

Periodic Comets of Only One Appearance

Comet	Perihelion Date	P	q	H_0	Remarks
Boethin	1975 Jan. 5	11.04	1.10	10.5	Infrequent favorable returns.
Sanguin	1977 Sept. 17	12.61	1.81	13.5	
Peters	1846 June 1	13.38	1.53	8.0	
Gehrels 1	1973 Jan. 25	14.54	2.94	11.5	
Kowal 1	1977 Mar. 3	15.13	4.66	9.0	
Van Houten	1961 Apr. 29	15.75	3.94	8.0	Discovered in 1964 on plates exposed in 1960.
Chernykh	1978 Feb. 14	15.93	2.57	7.0	
Pons-Gambart	1827 June 7	63.83	0.81	7.0	
Dubiago	1921 May 5	67.01	1.12	10.5	
De Vico	1846 Mar. 6	75.71	0.66	7.2	
Vaisala 2	1942 Feb. 15	85.42	1.29	13.2	
Swift-Tuttle	1862 Aug. 23	119.98	0.96	4.0	Bright comet; responsible for Perseid meteor shower.
Barnard 2	1889 June 21	145.35	1.11	9.0	
Mellish	1917 Apr. 11	145.36	0.19	7.4	Fairly bright comet.

LOST PERIODIC COMETS

Short-period comets may become "lost" for a number of reasons, not the least of which being poor observability and the consequent inability of astronomers to determine a sufficiently accurate orbit to facilitate recovery of the comet at a subsequent return. This is especially critical where the comet passes close to Jupiter and may, in consequence, have its orbit drastically altered. In such circumstances, only accurate determination of the orbit will allow recovery and, where this is not possible, the comet will most probably become "lost."

Some comets (e.g., P/Gale and P/Boethin) are so unfavorably situated that they can only be observed from Earth during those years in which their perihelia fall within a narrow "window" of dates. The chances of this occurring are not very great, especially when the period is an integral multiple of one year, and such comets may easily become "lost."

It is surely a demonstration of both the power of modern methods of computation and the concerted efforts of observational astronomers using large telescopes that a number of traditionally "lost" comets have been recovered in recent years (P/Holmes, P/Taylor, etc., in addition to those accidentally rediscovered—e.g., P/Perrine-Mrkos, P/Denning-Fujikawa, and P/Swift-Gehrels); and this fact further confirms the contention that it is the lack of an adequate ephemeris which leads to the loss of most comets and not a rapid fading of the comets themselves. If a better ephemeris becomes available, the comet is frequently recovered.

Certain comets, however, do appear to be lost forever. P/Lexell and P/Oterma have been perturbed into larger orbits and no longer approach Earth sufficiently closely to be observed, although there is no serious question concerning the continued existence of these objects. Nevertheless, others appear to have truly faded and probably disintegrated. The strongest candidates for this group are P/Brorsen, P/Biela, P/Westphal, and probably P/Perrine-Mrkos. P/Haneda-Campos also appears unstable and may soon become lost.

Table Three

Daylight Comets

Included here are those comets, and objects which may have been comets, which have been observed in full daylight.

For the purposes of definition, we shall use the term "daylight comet" to include all those observed, with or without optical aid, in light of visual wavelength, by an observer located at or near the surface of the Earth at a time when the uneclipsed Sun was above the horizon. Thus objects observed from artificial Earth satellites, or at infrared wavelengths during daytime hours, or in very bright twilight or during total solar eclipses, have been excluded.

Thus this list includes the brightest comets on record, as well as others seen under exceptional conditions.

Comet	*Remarks*
1. 146 (May) B.C.	Chinese records mention that it "moved away at dawn and became smaller," *possibly* indicating visibility after sunrise.
2. 43 B.C.	The "Julian Comet."
3. 302 A.D.	Comet seen in daylight during May or June, according to Oriental records.
4. 363	Vague mention of daylight comets "during reign of Jovian" or "toward end of year," in Roman chronicles. Uncertain connection with the August–September comet of that year.
5. 368 (?)	Number 4 misplaced?
6. 437	Large yellow star seen in day according to Chinese records. Exact nature of

	object not clear.
7. 520	Comet in October and November said to have been visible "in the morning" on November 30. The ground for suspecting daylight visibility rests upon the assumption that the word translated "morning" in Chambers' catalogue is the same as that similarly translated in the catalogue with reference to the comet of 302. This latter was translated as "daylight" by Ho Ping-Yu.
8. 575 (?)	Possible daylight comet this year according to Chambers. The object seen in April of that year may have been a nova.
9. 1077 (?)	Old English record tells of a "blazing Starre seen near unto the Sonne" on Palm Sunday. Pingre suggested that this may have been Venus near inferior conjunction, but the term "blazing star" was applied to comets and this possibility remains open.
10. 1106, Feb. 4 or 5	European records report a star or comet "one foot and a half" (i.e., presumably, 1–2 degrees) from the Sun, and later a bright comet was seen in the southwest after sunset. Probably a Sun-grazer. Magnitude probably about —10.
11. 1222, Sept. 9	Comet P/Halley seen according to Oriental records in daylight on this day.
12. 1264	Very large comet. Chinese records may be interpreted as implying visibility after sunrise for about a month. Orbit has been determined and suggests a

Comet	*Remarks*
	perihelion distance of about 0.8 A.U. and an absolute magnitude possibly as high as 2.
13. 1382	Korean records report a star seen on September 5 and an "auspicious star" seen during the night. Daylight visibility seems implied here, but the exact nature of the "star[s]" is not clear.
14. 1402	One of the finest comets on record, seen in February. Visible in daylight for eight days—a record if Comet 1264 is excluded. Magnitude about —5. Rough orbit calculation suggests perihelion of about 0.4 A.U. and absolute magnitude of around +0.5, but these figures are not accurate. Nevertheless, the great brightness of the comet seems to have been intrinsic rather than due to a close approach to the Sun.
15. 1402	Chambers records a large comet in August, said to have been visible in daylight. (No. 14 misplaced?)
16. 1472	Great comet recorded in China as visible "even at midday." Perihelion 0.49 A.U. and absolute magnitude about 2. Close approach to Earth.
17. 1532	Chambers records a daylight comet this year—presumably comet of September and October. Probable magnitude about —2 or —3. Perihelion distance 0.52 A.U. Absolute magnitude about 2.

18. 1577 — Tycho's Comet. About magnitude —4 at discovery, with head visible in daylight. Perihelion 0.18 A.U. Absolute magnitude around —1.

19. 1587 Aug. 30 — Japanese records report an object like a "guest star" visible throughout the day. Association with other comets seen that year is unclear, and exact nature of the object is not known. (Could it have been Venus or one of the bright stars or planets seen in an unusually clear sky?)

20. 1618 II — The great comet of that year is recorded by Chambers as having been visible in daylight. Perihelion distance 0.39 A.U. Absolute magnitude 4.6.

21. 1731 I, Feb. 9 — Seen "in sunlight" from Lisbon and in Gibraltar (magnitude —1 or —2?). Perihelion distance 0.22 A.U. Absolute magnitude 4.0.

22. 1744 — De Cheseaux's Comet seen in daylight late February about 12 degrees from Sun. About magnitude —5. Perihelion distance 0.22 A.U. Absolute magnitude +0.5.

23. 1843 I — Brightest comet since 1106, at that time. Conspicuous naked-eye object within a few degrees of the Sun. At least 3 degrees of tail visible in daylight with the naked eye. Magnitude must have been about —8, —10.

24. 1847 I — Hind's Comet, observed by Hind

Comet	*Remarks*
	(March 30) in daylight near perihelion. Seen with the aid of a telescope and a bright green filter. Magnitude probably about —3, —4. Drawing made. Perihelion distance 0.04 A.U. Absolute magnitude 6.8.
25. 1853 III	Klinkerfues' Comet. Observed telescopically in daylight for several days late August and early September, when located about 8 degrees from the Sun. Magnitude about —1. Perihelion distance 0.31 A.U. Absolute magnitude 4.4.
26. 1858 VI	Rev. T. W. Webb notes that Donati's Comet was observed in daylight by its discoverer on October 8, but only with the aid of a large and powerful telescope. Magnitude about 0.0. Perihelion distance 0.58. Absolute magnitude 3.3.
27. 1861 II	Tebbutt's Comet seen in daylight, early in July, as a starlike object of about magnitude —3. Perihelion distance 0.82. Absolute magnitude 3.9. Close approach to Earth.
28. 1882 I	Wells's Comet observed telescopically and spectroscopically when very near the Sun. Brilliant lines of sodium seen in spectrum even after sunrise. Magnitude about —4. No tail seen in daylight. Perihelion distance 0.06 A.U. Absolute magnitude 4.1.

29. 1882 II — One of the most brilliant comets on record, was observed to the limb of the Sun on September 17, when it was at least magnitude —10. About ½ degree of tail seen with naked eye in daylight.

30. 1882 — In December, a starlike object was seen near the Sun by several people in Scotland. No other reports of this object are known, and its cometary nature remains problematical.

31. 1896 — On September 20, L. Swift reported a small comet 1 degree from the Sun (presumably in the daytime). The next day, he reported that it had moved north and become fainter. (This northward motion would seem to preclude any association with the Sun-grazing group of comets.)

32. 1901 I — Viscara's Comet observed telescopically for a period of some fifteen minutes after sunrise at Yerkes, on April 24, when located 15 degrees north of the Sun. Magnitude —1. Only "nucleus" could be seen. Perihelion distance 0.25 A.U. Absolute magnitude 5.9.

33. 1910 I — Brilliant comet seen in daylight during mid-January. It was a naked-eye object when within 4 degrees of the Sun, and when observed from suitable localities, revealed a naked-eye tail of 1 degree in daylight. Judging by recorded accounts, it must have been about magnitude —5 or —6. Perihelion distance 0.13 A.U. Absolute magnitude 5.0.

Comet	*Remarks*
34. 1921e, Aug. 7	A bright starlike object (possibly magnitude —4 or a little brighter) was observed with the naked eye near the setting Sun by a group of Lick Observatory astronomers, including H. N. Russell and W. W. Campbell. A number of other possible sightings were reported about this time, but the exact nature of the object could not be determined, although the cometary explanation seems most reasonable. Peculiar sky effects were also allegedly reported about the same time, and some people believed these to be due to the Earth's passage through a comet's tail, although this is doubtful.
35. 1927 IX	Skjellerup's Comet. Brilliant comet observed with the naked eye when only 1.5 degrees from the Sun on December 15, 16, and 17. Still a daytime object on December 20, and a stubby tail was visible through a comet-seeker only five minutes before sunrise on December 21. At its brightest, the comet seems to have been about —6 or —7 and to have shown a ½-degree tail (visible to the naked eye) in daylight. Daytime photograph obtained on December 15 at Lowell Observatory. Perihelion distance 0.17 A.U. Absolute magnitude 5.2.
36. 1965 VIII	Ikeya-Seki Comet. Together with comets of 1106, 1843, and 1882, this was one of the most brilliant on record. Seen with naked eye within 1 degree of

	Sun at about magnitude —10 or —11. Some 2¼ degrees of tail visible with the naked eye in daylight—tail said to have been as bright as the crescent Moon. Photographs and spectrograms obtained. Comet observed through perihelion passage, when brightness probably exceeded —13.
37. 1976 VI	West's Comet. Observed with the aid of binoculars and small telescopes in daylight when near perihelion. Observed with the naked eye by John Bortle ten minutes before sunset on February 25, 1976. Magnitude estimated as —3. Perihelion distance 0.20 A.U. Absolute magnitude 5.0.

APPENDIX FOUR
MAGNITUDES OF ASTRONOMICAL OBJECTS

The term "magnitude" has been used frequently throughout this book as a measure of the brightness of an astronomical object (see the Glossary for details).

Listed below are a number of objects in terms of decreasing brightness, which should help the newcomer to astronomy to form some idea of the magnitude scale.

Object	*Magnitude*
Sun	—26 (approx.)
Full Moon	—12.5
Quarter Moon	—10
Venus at maximum brightness	—4.4
Jupiter at maximum brightness	—2.8
Sirius	—1.43
Vega	+0.04
Regulus	1.36
Alpha Hydrae	2.16
Zeta Herculis	3.0
Delta Ceti	4.04
Andromeda Nebula	4.8
Epsilon Trianguli	5.44
Faintest star seen with naked eye on good night	6 (approx.)

Faintest star seen with 7 × 50 binoculars on good night	9 or 10
Faintest star photographed with very large telescope	22 (approx.)
Faintest star photographed with largest telescopes	24 (approx.)
Representative comets are:	
1882 II and 1965 VIII	about —10 to —17
Typical naked-eye daylight comet	about —5 to —8
West of 1976	—3
Bennett of 1970	0
Arend-Roland of 1957	1
Ikeya of 1963	2 or 3
Tago-Sato-Kosaka of 1969	3 or 4
"Normal" long-period comet	7 to 10
"Normal" short-period comet	15 to 18

GLOSSARY

The following glossary contains the more frequent terms used in this book. It is not intended to be a complete list of all terms employed in cometary astronomy.

Aphelion: The point in the orbit of a planet or comet at which the object is farthest from the Sun.

Bolide: A bright fireball or meteor, especially one which explodes.

Coma: The luminous region of a comet surrounding the nucleus.

Desorption: The release of molecules of gas, concentrated in a porous substance, with increasing heat. The opposite process (the "taking up" of gas molecules with decreasing temperature) is known as *adsorption.*

Ellipse: A closed curve. The shape of the orbits of objects in revolution around the Sun.

Hyperbola: An open curve. An object moving in an hyperbolic orbit around the Sun would theoretically escape to infinity.

Ices: Frozen gases or liquids which melt or evaporate at moderate or low temperatures.

Magnitude: The *apparent* magnitude is a measurement of how bright an object appears to the eye of an observer. An object of one magnitude is approximately 2.51 times brighter than an object of the next magnitude, and the smaller values indicate brighter objects. Thus an object of first magnitude is 2.51 times brighter than an object of second magnitude, and an object of magnitude —1 is $(2.51)^2$ or 6.3 times brighter than one of first magnitude.

The *absolute* magnitude (of a comet) is the hypothetical apparent magnitude which the object would have at one astronomical unit from both Earth and Sun.

Meteor: A track of light in the sky from rock or dust burning up in the Earth's atmosphere.

Meteorite: An interplanetary chunk of rock after it impacts on a planet or a moon, especially on the Earth.

Meteoroid: An interplanetary chunk of rock smaller than an asteroid.

Nucleus (of a comet): The most solid part of a comet, containing nearly all the cometary mass.

Occultation: The hiding of one astronomical body by another.

Parabola: An open curve. Theoretically, a body moving along a parabolic path will reach infinity, but will arrive with zero velocity after an infinite lapse of time.

Perihelion: The near point to the Sun of the orbit of a body orbiting the Sun.

Perturbation: The disturbance in the motion of an astronomical body caused by the gravitational attraction of another. *Planetary perturbations* are vital in determining the true paths of comets within the solar planetary system. *Stellar perturbations* are believed responsible for the deflection of comets from the circumsolar cloud into the inner solar system, on the one hand, and into interstellar space on the other.

Return: A term used for the reappearance of a comet at a specific perihelion passage.

Tail: A nebulous appendage, composed of either gas or dust, associated with some comets.

INDEX